THE DEVELOPMENT OF BRAIN
AND BEHAVIOUR IN THE CHICKEN

Professor Lesley Rogers holds a personal chair in the Department of Physiology at the University of New England, Australia. She obtained her first degree in Zoology from the University of Adelaide and completed a D.Phil. at the University of Sussex in England under the tutelage of Professor Richard J. Andrew. In 1987 she was awarded a Doctorate of Science at the University of Sussex on the basis of her many years of research into the development of brain and behaviour in the chicken. She has published widely in leading scientific journals as well as for a more general audience. Her previous book written jointly with Professor John L. Bradshaw is entitled *The Evolution of Lateral Asymmetries, Language, Tool Use, and Intellect*, Academic Press, San Diego.

To the memory of my parents, Joyce and Owen Rogers

THE DEVELOPMENT OF BRAIN AND BEHAVIOUR IN THE CHICKEN

Lesley J. Rogers, D.Phil., D.Sc.

Department of Physiology
University of New England
Armidale
NSW 2350
Australia

CAB INTERNATIONAL

CAB INTERNATIONAL   Tel: +44 (0)1491 832111
Wallingford   Telex: 847964 (COMAGG G)
Oxon OX10 8DE   E-mail: cabi@cabi.org
UK   Fax: +44 (0)1491 833508

A catalogue entry for this book is available from the British
Library.

ISBN 0 85198 924 1

Typeset in Photina and Optima by Solidus (Bristol) Limited
Printed and bound in the UK at The University Press, Cambridge

CONTENTS

PREFACE

As a model for studying development, the chick has provided valuable insight into broad issues of development. The findings have had an impact on general theories of development, even as they relate to the human species. The now well-documented and complex interactions between genetic, hormonal and environmental factors demonstrated in the developing chick will be recognized as providing a potent argument against unitary causal explanations for differences in behaviour. Given the resurgence, in scientific circles as well as in popular scientific writing, of genetic or hormonal explanations for the cause of differences in human behaviour, it seemed timely for me to emphasize the interactive processes of development and to illustrate its complexity 'even in the chicken'.

In this book my aim has been to include important old and new research in the fields of brain development and behaviour in the chick, and to juxtapose them to similar work with other avian species and, to a lesser extent, mammalian species. I believe that a full understanding of development can not be achieved without considering the integrated associations between behaviour and brain development at both the cellular and subcellular levels. Hopefully, this book will make these relationships clear to the reader. Much more could be done to draw out and expand on the topics presented. I wanted to raise questions, and sometimes, but not always, to answer them.

Chapters 1 and 2 outline the developmental stages of the chick embryo, with emphasis on the effects of environmental stimulation in Chapter 2. Readers more interested in behaviour might begin at Chapter 2. Posthatching behaviour is discussed in Chapter 3, and followed in Chapter 4 by the neurochemistry of development and memory formation in the posthatching period. Chapter 5 presents the rather remarkable transitions in behaviour and neurochemistry that occur during the first two to three weeks of posthatching life and finds the best explanation for these in terms of changing hemispheric dominance. Chapter 6 attempts to place the research with chickens into the wider context of avian and mammalian species, although some references to this are made throughout the book. The final chapter deals with the new and exciting field of avian cognition and moves on to discuss some issues of welfare for the domestic chick.

I wish to thank Professor Richard Andrew and Dr Bryan Jones for their valuable comments and suggestions for improving this manuscript. Indeed, I am indebted to Professor Andrew's contribution to my thinking throughout my scientific career. There were also anonymous reviewers of this manuscript, whom I thank for helpful and encouraging comments.

I am most grateful for the untiring help with library material and manuscript preparation given to me by Gavin Krebs and to Helene Dawson for her ever willing help with typing and other support. I also thank Michele Grey for helping to type the manuscript and John Roberts for his excellent photography and preparation of many of the figures, Ivan Thornton for producing some of the figures, especially the cover illustration, Amy Johnston and Tom Burne for useful comments on some of the chapters and particularly Dr Gisela Kaplan, who gave much valued help in proof reading and indexing. In a more general sense, I am grateful to my colleagues, students and friends who shared ideas or contributed support for this project.

— 1 —

DEVELOPMENT OF BRAIN AND BEHAVIOUR BEFORE HATCHING

Summary

- This chapter discusses the development of brain and behaviour occurring from the beginning of incubation of the eggs until hatching.
- The central nervous system begins to develop on the second day of incubation and reaches a comparatively well-developed form prior to hatching. Its development and the known changes in the sensory systems are discussed, together with correlated changes in behaviour.
- Information on the neurophysiology and function of each sensory system is given, with particular detail for the visual and auditory systems.
- The sensory systems become functional according to the sequence: tactile (day E6), proprioceptive/vestibular (day E8 to E10), auditory (day E12 to day E14), taste (not definitely known, but possibly about day E12), visual (day E18), and olfaction (probably on day E20).
- Each of the sensory systems develops according to its own programme although, as shown in Chapter 2, there may be interactions between the various sensory systems that deserve consideration at the anatomical and neurochemical level in future research.
- Within the visual system, the different pathways become functional at different stages of development. Development of the tectofugal, thalamofugal and isthmo-optic pathways is outlined. It appears that the tectofugal visual system becomes functional in advance of the thalamofugal visual system.

Introduction

Encapsulated in its shell and membranes the chick embryo undergoes sufficient development to hatch out as a relatively independent individual capable of feeding and drinking alone and ready to learn behavioural associations essential to survival. In these respects the chicken (*Gallus gallus* or *Gallus gallus spadiceus*, the jungle fowl, and *Gallus gallus domesticus*,

Fig. 1.1. The embryo in the egg interacts with four interconnected environments. The first (**1**) is the internal environment of the embryo itself, the second (**2**) that inside the egg, the third (**3**) includes the environment inside the nest and the hen and the fourth (**4**) is external to all of these. This figure has been adapted from Freeman and Vince (1974).

referring to the domesticated strains) differs from altricial species, which after they hatch require considerable parental care together with confinement in the nest. It is the chicken's precocious stage of development at hatching that appears to have led to its choice as a species for domestication, along with other Galliformes (turkeys and quail) and Anseriformes (ducks and geese). The majority of mammals that have been domesticated are also precocial; for example, horses, goats and sheep (Hale, 1969; Wood-Gush, 1983). In 21 days of incubation the chick embryo must achieve a remarkable degree of structural and functional development. This requires precisely timed sequences of development in all of the growing tissues and organs, including the brain.

It might be a relatively simple matter to read out a genetic programme for development with speed and precision, even though such a programme would require complex encoding in the genome, but the programme for development is played out in interaction with the environment surrounding the embryo and also external to the egg itself. Differential effects of influences from the environment on the growth processes must be regulated or coordinated into the overall scheme of development.

The shell and membranes of the egg offer only a partial shield of the embryo from the environment. Indeed, gaseous exchange as well as temperature and humidity balances must occur across these barriers. As a simple model we may consider the embryo as developing amid four intercommunicating environments (Fig. 1.1). The first is the internal environment of the embryo itself, which includes cellular interactions, circulating hormones, internal oxygen supplies and much more. The second environment exists between the embryo and the eggshell, and includes the various surrounding membranes, the yolk sac and the albumen. It is this environment that provides the embryo with sustenance and gas exchange while cushioning it from many of the changes occurring external to the egg. The third environment extends from the eggshell to the nest cup, an environment which largely serves to insulate the egg(s) from temperature variations and one which includes the hen herself. The hen's function in this environment is to provide warmth, to turn the eggs and to provide vocal stimulation, thus allowing the embryos to form attachments which will ensure their survival after hatching (see later). The fourth environment is that of the external world which must inevitably, as well as essentially, communicate with all of the other environments. For example, as will be discussed in some detail, light present in the fourth environment readily penetrates to the embryo, provided the hen is off the nest, reaching it by passing through the egg shell and membranes. Light reaching the embryo influences the rate and pattern of its development. Not only does light affect growth of the embryo in general, but it has particular effects on the developing central nervous system. Primarily, it affects the development of the visual pathways, but this has repercussions on the other sensory systems and consequently it determines the responses of the hatched chicken to stimuli in the various modalities. The developing brain is in fact a focus for the interaction between genetic and environmental influences. Moreover, the brain of the embryo, during the later stages of incubation at least, can learn and so attachments can be formed to stimuli present in environments two, three and four. Therefore, just as in mammals, the environmental influences do not begin at birth or hatching. Rather, they are interacting with genetic determinants throughout embryonic and postnatal or posthatching life.

Towards the end of the incubation period the embryo itself can exert some control over environment three by communicating with the hen. For example, certain vocalizations made by the chick embryo stimulate the hen to turn the eggs and to vocalize in turn (Tuculescu and Griswold, 1983, and see later). The results of this interaction are tactile and auditory stimulation of the embryo, together with the regulation of temperature. Stimulation of the embryo by changing intensity of light as the hen stands up and resettles herself on the nest may also occur.

After hatching, for precocial species like the chicken, environments two and three disappear, whereas for altricial species environment three persists until fledging. *Gallus gallus* chicks do spend most of their time in the nest for the first day after hatching, but not longer than this. Thus, hatching for a precocial species means a more dramatic change in environmental impact than it does for altricial species. Consequently, dramatic changes must occur in preparation for the hatching process itself as well as for independent life immediately after hatching. As we shall see, these changes are manifest in the various aspects of brain development and its function.

Development of the Central Nervous System and Activity Patterns

The brain begins to differentiate as early as the second day of incubation (E2; the E refers to embryonic age as distinct from age after hatching). Its development continues throughout the incubation period, during which time

Fig. 1.2. The brain and eyes of chick embryos at day E10 (left) and E20 (right) of incubation. Note the increase in brain size over these two stages of embryonic development, particularly of the forebrain and cerebellum, and the increase in the size of the eyes.

its size increases dramatically (Fig. 1.2). The development continues after hatching, particularly during the first two or three weeks of independent life. In so doing, it passes through a series of precisely timed developmental events, which are manifest in rather discrete transitions in behaviour of the embryo and chick. Compared to slower developing species, certain neurochemical and structural developments occur ubiquitously and simultaneously in the chicken forebrain. For this reason, the domestic chicken is frequently used in studies of brain growth and maturation (e.g. Rostas, 1991) and to investigate the effects of neurotoxins on development (e.g. Sanderson and Rogers, 1981). For example, drugs which interfere with neurogenesis are likely to have a maximal affect when they are administered to the embryo on day E8 of incubation because this is the peak period of neuron formation in the developing forebrain (Freeman and Vince, 1974; Sobue and Nakajima, 1978). The formation of synaptic connections between neurons in the forebrain peaks later, at day E15 of incubation (Sedláček, 1972). Although some of these synaptic connections may immediately begin to mature, which means taking on functions characteristic of synapses in the adult brain, most do not mature until well after hatching, between days ten and 60 posthatching (Rostas, 1991, and Chapter 4). Thus, the proliferation and maturation phases are largely distinct from each other. By contrast, in the slower developing rat brain the periods of synapse proliferation and maturation overlap, the former occurring from birth to approximately day 25 and the latter from about day five to day 30 (Markus *et al.*, 1987).

As a general pattern, neuromotor and neurosensory development involves four distinct, but overlapping, processes: proliferation, migration, differentiation and synapse formation. In the chick embryo these processes are relatively well separated from each other.

The main stages of nervous system development are listed in Table 1.1. The spinal cord and the peripheral nervous system differentiate during the first week of incubation so that by day E7 the spinal tracts are well developed. Beginning as early as day E3 or E4, the chick embryo displays spontaneous rhythmical movements of the head and body (Hamburger and Balaban, 1963). The activity begins with movement of the head and neck, and, as development proceeds, it includes the trunk and then the tail region. Active movements of the limbs do not commence until day E6, at the same stage when division of the motor neuron cells has been completed (Foelix and Oppenheim, 1973). The limbs do not move independently until day E7. Active periods of 5–15 s duration alternate with inactive periods of 30–60 s at a rather constant rhythm until day E8, when the duration of the activity phases increases. From day E12 until just before hatching (after the embryo has pipped the egg shell), the activity of the embryo is continuous (Hamburger, 1963), and it peaks at day E13 (Foelix and Oppenheim, 1973).

Electrophysiological recording from the spinal cord has revealed the presence of polyneuronal burst activity beginning on day E5. In parallel with

Table 1.1. A chronological chart of the development of the various sensory systems, hormonal secretion and general behaviour of the embryo. The age of the embryo is listed as day of incubation. It might be argued that the stage of development should be given according to the much cited, descriptive list of developmental stages of chick embyros compiled by Hamburger and Hamilton (1951, and reprinted in 1992), but for most of the information included the authors cite developmental age in terms of day only, and many events are not precisely timed even to within a single day. References to the various developmental events listed can be found in the text.

Day	Brain and spinal cord	Visual	Auditory	Olfactory/taste	Tactile, proprioceptive and vestibular	Hormones	General behaviour of embryo
1	Neural folds appear.						
2	Brain differentiation begins. Neurohypophysis begins to form. Neural tube begins to close.	Optic vesicles differentiate. Axons from retinal ganglion cells begin to leave retina.	Inner ear begins to develop.	Nasal placode forms.	Neurogenesis in vestibular ganglion begins.		
3		Pigmentation of eyes begins. Amacrine cells begin to form in retina. Light exposure increases activity of embryo (d 3–6).	8th nerve begins to grow into the brain. Neurogenesis of the nucleus magnocellularis.		Sensory fibres from facial ganglion begin to enter the brain.		Passive movements detected. Flexion of neck. Head turns to left.

4	Most of the basal forebrain structures have formed.	Green light increases duration of movement, blue and red without effect.	Neurogenesis in cochlear ganglion begins. Cochlear nucleus visible. Neurogenesis of the nucleus laminaris.	Nasal pits have differentiated. Olfactory nerves grow towards brain. Olfactory bulb has formed.		Synthesis of oestrogen and 17β-oestradiol begins.	Active movements of head and neck. Embryo turns to left.
5	First synapses can be detected in the spinal cord. Polyneuronal burst activity first recorded from spinal cord.	Neurogenesis of isthmo-optic nuclei. Ectostriatum has formed.	Oronasal grooves form. Other olfactory regions of the brain have formed (olfactory tuberculum and piriform cortex).		End of neurogenesis in vestibular ganglion. Nucleus basalis (somatosensory) and archistriatum ventrale (somatomotor) areas have formed.	Synthesis of corticosteroid hormones begins. Synthesis of testosterone begins. LH-RH neurons present in hypothalamus.	Active movement of trunk. Mouth movements.

Table 1.1. *continued*

Day	Brain and spinal cord	Visual	Auditory	Olfactory/taste	Tactile, proprioceptive and vestibular	Hormones	General behaviour of embryo
6	Most motor neuron cell formation has been completed. 50% of the neurons of the neostriatum have formed.	The neurotransmitters GABA and acetylcholine are present in the retina. Axons from retina arrive at optic tectum. Peak neurogenesis of most retinal cells. Green light exposure decreases activity (d 6–8), red light increases activity.	8th nerve enters the nucleus magnocellularis (one of the brain stem auditory nuclei).	Upper beak has grown between the nostrils. Nasal capsule begins to form. The olfactory epithelium is present within the nasal capsule.	Tactile sensitivity.		Passive movements cease. Limbs participate in whole body movements.

7	Marked development of spinal tracts. Rhythmic activity of motor neurons first recorded. 50% of the neurons of the hyperstriatum ventrale (in forebrain) have formed.	First eyelid movements detected.	Neurogenesis in cochlear ganglion ends.	Olfactory neurons have reached the olfactory bulb.		Testosterone secretion begins to increase, and more so in males. Plasma 17β-oestradiol begins to increase.	Eyelids move. Independent limb movements.
8	Neurogenesis peaks. 50% of the neurons of the hyperstriatum accessorium have formed.	First eyeball movements detected.	Hair cells appear in cochlea. Field L (primary auditory area of the forebrain) begins to form.		Vestibular sensitivity.	ACTH production begins. Adrenal gland secretes adrenaline.	Movement of eyeballs.
9	Cholinergic neurons present in spinal cord.	Amacrine cell generation is completed. Branching and arborization of neurons in the optic tectum begins. Glial cells begin to form in the optic tectum.	Projections from acoustic ganglion enter cochlear nuclei.				

Table 1.1. *continued*

Day	Brain and spinal cord	Visual	Auditory	Olfactory/taste	Tactile, proprioceptive and vestibular	Hormones	General behaviour of embryo
10	Synapses in the spinal cord can be fully identified. Glial cell number in the telencephalon increases.	Cell death begins in accessory oculomotor nuclei. Axons from isthmo-optic nuclei begin to reach retina. Cholinergic cells in retina are fully developed. Synaptogenesis of retinal cells begins.		Digestive processes commence.	Local proprioceptive muscle reflexes.	Thyroid hormone secretion. 17β-oestradiol first binds to cells in preoptic area of hypothalamus.	Body movements become jerky and random.
11	First synapses can be identified in the cerebellum. Myelination in spinal cord and of peripheral nerves begins.	Bipolar cells of retina begin to differentiate. Isthmo-optic nucleus has mature form.	Myelination of 8th nerve begins. Auditory evoked responses from brain stem auditory nuclei. Cell death in auditory nuclei begins.				

12	Glutathione peroxidase levels begin to increase.	Cell death begins in isthmo-optic nuclei. Rapid growth of cells in optic tectum begins plus formation of functional synaptic contacts.	Movement occurs in response to low frequency auditory stimulation.	Imbibition of amniotic fluid.	Marked increase in the number of luteinizing hormone (LH) producing cells.	Continuous activity begins – it continues to be unintegrated random movements.
13	Polyneuronal burst potentials peak. Local inhibitory mechanisms mature. EEG detected.	Differentiation of nerve cells in the retina is completed. Synaptogenesis proceeds in the inner plexiform layer of the retina. Axons from isthmo-optic nuclei form synapses in retina.	Cell death in brain stem auditory nuclei is completed. Brain stem elements of auditory system functional. Cochlear microphonics recorded.		Pituitary–gonadal axis definitely established. Neurohypophysis active. In females, 17β-oestradiol secretion begins to increase more rapidly. In males, plasma testosterone levels peak.	Activity peaks.

Table 1.1. *continued*

Day	Brain and spinal cord	Visual	Auditory	Olfactory/taste	Tactile, proprioceptive and vestibular	Hormones	General behaviour of embryo
14	Cholinergic neurotransmission in spinal cord well developed. Calcium-binding protein in spinal neurons has reached a peak.	Axons to isthmo-optic nucleus from tectum form synapses. Low amplitude activity in the optic tectum. Glial cells form in the inner retina and myelination of the central retina begins.	Activity decreases in response to 700 and 1400 Hz tones.			Females begin to produce oestriol.	Beak clapping begins at approximately this stage (E12–16), but the sound produced is probably inaudible.
15	Synapse formation peaks. GABA neurotransmission starts to develop.	Myelination of optic fibres begins. Neurally mediated pupillary reflex.	Cells in auditory nuclei increase their connections and size. Cochlea is mature. Auditory thresholds begin to mature and continue to do so until day 1 posthatching.			Thyroid activity increases. Plasma testosterone levels in females peak.	

Day						
16	Stratification of cerebellar cortex is visible. Electrical activity in cerebellum.	40% of retinal ganglion cells die. End of period of synaptogenesis in the inner plexiform layer of the retina. Axonal arborization in optic tectum complete. Retinotectal system mature.	Activity increases in response to 700 and 1400 Hz tones. Some capability for auditory conditioning.			Capable of respiratory movements.
17	Slow waves alternate with faster low amplitude waves in forebrain (sleep?). Number of Ca^{2+} ion channels begins to increase and rises rapidly to day E21. Purkinje cells in cerebellum are established.	Electrical activity recorded from retina. Visually evoked responses recorded from optic tecta. Isthmo-optic system mature (cell death completed).	Activity increases in response to 400, 700 and 1400 Hz tones.	Tactile conditioning is possible.	Plasma luteinizing hormone levels equal those of adult. In males, plasma testosterone levels have declined to a trough. In females, 17β-oestradiol levels begin to fall slightly until after hatching.	Coordinated stereotyped movements commence. Eye opening and closing. Stable habituation possible. Unstable conditioned reflexes.

Table 1.1. *continued*

Day	Brain and spinal cord	Visual	Auditory	Olfactory/taste	Tactile, proprioceptive and vestibular	Hormones	General behaviour of embryo
18	GABA receptor numbers in forebrain have reached a maximum. Increased rate of dendritic arborization and increased GABA synthesizing enzyme in cerebellum.	Electroretinogram occurs. Electrical activity in optic tectum is mature. Behavioural response to light (opening and closing of eyelid, beak clapping, vocalizing). Discrimination of red versus blue.					Tucking occurs.
19	EEG pattern matures. Paradoxical sleep. Forebrain slow waves attain regularity. Sleep patterns occur. Purkinje cells in cerebellum increase in size.		Reduced auditory threshold (fluid clearance from middle ear). 8th nerve myelination complete: increased conduction velocity.			Plasma levels of LH peak.	Postural reflexes fully developed. Vocal apparatus is functional. Embryo in the hatching position.

20	Na+ and Ca2+ ion channels increase markedly. Metabolic activity of hyperstriatum decreases.	Light-evoked potentials in forebrain.	Improved ability for auditory conditioning. Responses to maternal calls.	Nares freed of obstructing materials. Olfactory discrimination.	In males, plasma testosterone levels have increased again. In females, they have risen consistently.	Beak penetrates air sac. Breathing begins. Clicking and vocalizing begin. Pipping of egg shell. Beak clapping increases.
21	Forebrain activity increases (but metabolic activity of HA and IMHV decreases).	Some visual regions still developing. Discrimination of red, blue and orange.				Stable conditioned reflexes. Clicking stops temporarily. Hatching.

Abbreviations: GABA, γ-aminobutyric acid; EEG, electroencephalogram; HA, hyperstriatum accessorium; IMHV, intermediate medial region of the hyperstriatum ventrale; LH-RH, luteinizing hormone-releasing hormone.

the level of motility, this burst activity peaks on day E13 and then declines to reach almost the same level on day E20 as it was on day E5 (Provine, 1972, 1973). Before day E13 the burst activity cycles between periods of low and high levels, as confirmed by O'Donovan and Landmesser (1987). On day E13 this regular periodicity of the burst activity is lost, suggesting to Provine (1973) that a threshold has been lowered on a cyclic process; that is, the electrical activity in the spinal cord may then be continuously above threshold and no longer constrained by self-limiting cyclic processes. Possibly the spinal cord comes under inhibitory influences which suppress spontaneous discharges. Along with these changes in electrical activity, motility of the embryo changes in a consistent manner. In fact, as would be expected, the electrical discharges are synchronous with movements of the embryo.

The motor activity occurring early in development is apparently spontaneous, rather than being a response to sensory stimulation (Hamburger and Balaban, 1963; Corner and Schadé, 1967). This means that, in general, the early motor movements of the embryo are autogenous and non-reflexive (Gottlieb, 1976). Even movements of the limbs may occur autonomously, as shown by Hamburger *et al.* (1966), who deafferented the region at the level of the leg on day E2 and scored motility of the embryo throughout incubation. These embryos had no sensory input from the limb to the spinal cord, but they showed no abnormality of leg motility. Thus, the movements may be determined by spontaneous discharge of motor neurons in patterns that may be essential to normal differentiation. Spontaneous patterns of discharge may also be generated by the developing brain, although Hamburger *et al.* (1965) have shown that influence from the brain increases the level of activity of the limbs but not its rhythm.

As development proceeds sensory pathways begin to assume active control of body movements. Sensory fibres, for example, from the facial ganglion begin to enter the brain on day E2 (von Bartheld *et al.*, 1992), and active movements of the head and neck are seen on day E3 or E4 (Table 1.1). The reflex arc from sensory input to motor output may, however, not be established until later, because tactile sensitivity is not apparent until day E6. At this stage of development the embryo will show withdrawal movements of the oral regions in response to being touched by a fine hair (Gottlieb, 1968). Local proprioceptive muscle reflexes do not appear until day E10 (Freeman and Vince, 1974). From this stage, the body movements are described as being jerky and random, until day E12 when continuous activity begins, and continues until day E17. On day E17 the embryo is capable of performing coordinated, stereotyped movements known as type III movements (Hamburger and Oppenheim, 1967). These movements appear to be mediated by the midbrain (Oppenheim, 1972a, 1973).

The presence of random movements from E10 to E17 would suggest that coordinating and integrating mechanisms of the central nervous system are not yet operating, at least fully (Hamburger and Oppenheim, 1967).

Although there are connections between the spinal cord and the pontine reticular formation of the day E4 embryonic brain and between the spinal cord and a number of other brain regions on day E5.5 (Okado and Oppenheim, 1985), they may not yet have assumed control of motor behaviour. Neurons in the central nervous system continue to differentiate over the entire period from E10 to E17. In the spinal cord cholinergic neurons first make their appearance on E9 (Thiriet *et al.*, 1992). Calcium-binding protein, which indicates the state of synapse maturation, increases to a peak level in spinal neurons on day E14 (Antal and Polgár, 1993).

The integrative capacities of the central nervous system emerge in a simple form by about day E15 to E17, when the embryo can form unstable conditioned reflexes. Sedláček (1964) demonstrated that conditioning to an auditory stimulus (a tone) coupled with an electric shock was possible from day E16 on. Hunt (1949) found a similar result to presentation of a bell from day E15 on, although he considered the stimulus to be vibratory rather than auditory. Conditioning to auditory stimuli is well established by day E20, when the embryo learns to identify the hen's calls (Gottlieb, 1968, and see later). Tactile conditioning is possible on day E17 (Gottlieb, 1968).

By day E13 an electroencephalogram (EEG) can be measured from the forebrain. Indeed, sporadic electrical bursting activity can be measured in the brain on day E14 (one day after burst activity in the spinal cord has peaked; Peters *et al.*, 1965), and from this stage the various neurotransmitter systems begin to develop. For example, GABAergic neurotransmission starts to develop in the forebrain on day E15 and continues beyond hatching into, at least, the first week of life (Abe and Matsuda, 1992).

Development of neurotransmission

The amino acid, γ-aminobutyric acid (GABA), is the main inhibitory neurotransmitter in the vertebrate brain. It is known to be widely distributed in the chick brain (Stewart *et al.*, 1988). In the avian brain GABA is known to play a role in feeding behaviour (Denbow, 1991), and in learning and memory (Chapter 4). One of its important neural functions is its inhibitory role in local circuits. These local inhibitory circuits are particularly important in the visual (Cuenod and Street, 1980) and auditory (Müller and Scheich, 1988) systems, where, for example, they provide lateral inhibition for directional selectivity for moving visual stimuli or changing sound frequencies (Heil *et al.*, 1992). GABA also appears to have a role in development and neural plasticity, particularly during synaptogenesis. There is some evidence that GABA promotes synaptogenesis (Wolff, 1981). Also, GABA receptor number is changed by visual experience during development (Fiszer de Plazas *et al.*, 1986) and by learning experience (Hyden *et al.*, 1984). The GABA agonist, muscimol, binds to GABA receptors and thus its tritiated form can be used to measure GABA receptor number and affinity. Muscimol binding in chick forebrain elevates markedly between day E12 and E18, from which stage it remains unchanged until day one posthatching when it begins to decline (Stewart and Bourne, 1986).

The level of GABA itself changes more than that of any other amino acid in the developing chick brain, increasing by fivefold from around day E8 to day one posthatching (Levi and Morisi, 1971). GABA levels in the brain may, of course, be influenced by many variables, including the level of enzymes involved in its synthesis from other amino acids and its uptake from the blood as the blood–brain barrier matures. The latter occurs during the second half of embryonic development (Bertossi *et al.*, 1993; Ribatti *et al.*, 1993) and the first week posthatching (Hambley and Rogers, 1979). The development of the GABA systems in the cerebellum seems to lag slightly behind those in the forebrain, with the enzyme for synthesizing GABA beginning to rise on day E17 or E18, a stage when synaptic proliferation in the cerebellum is also accelerating (Kuriyama *et al.*, 1968; Roberts and Kuriyama, 1968).

GABA has been discussed in preference to other neurotransmitters because it is widespread and its level increases so markedly during embryonic development. Glutamate is another amino acid neurotransmitter with a ubiquitous distribution in the vertebrate brain and with an important role in brain development (Fagg and Foster, 1983). It is an excitatory neuro-transmitter, which participates in mechanisms of development and neural plasticity (for reviews see Monaghan *et al.*, 1989, and Rogers, 1994). Glutamate receptors are involved in early memory formation after hatching (Chapter 4). It appears to be an important neurotransmitter in the visual pathways, as indicated by its presence in the optic tectum, together with GABA (Voukelatou *et al.*, 1992). The role of glutamate and receptors for glutamate deserves much more attention in the embryonic stages of development.

The release of neurotransmitters depends, in part, on the presence of calcium ion channels, and it is not until day E17 that these increase in number. From then on, they continue to rise sharply until hatching (Azimi-Zonooz and Litzinger, 1992). Thus, the last three to four days of embryonic development is the period when the structure and function of the central nervous system matures sufficiently to control all the behaviours necessary for hatching and neonatal life. Perhaps not surprisingly, the EEG takes on an essentially mature form by day E19 or E20 (Corner *et al.*, 1973).

The different rates of development of the various sensory systems will be discussed in sections to follow. First, it is worth considering the ideas motivating much of the early research on motility in the chick embryo that we have already covered.

Autonomous motor control and self-stimulation

Particularly in the 1960s a number of leading researchers were interested in the movement patterns of developing chick embryos because they were seen to provide evidence against the pervasive behaviourist position of the time; namely, that all behaviour is based on conditioned reflexes involving the stimulus–response link (Kuo, 1976). The findings of Hamburger and Oppenheim in particular (see p. 16) demonstrated the occurrence in embryos of autonomous motor behaviours that are not associated with sensory input. For example, as shown by Oppenheim (1968), at least up to day E15, overt cyclic motility of the chick embryo is not determined by sensory input from any source, although the direction and extent of these movements can be affected by sensory input. According to Hamburger *et al.* (1965), the reflex arcs for exteroceptive sensitivity have developed by day E7, and from this stage on they may control activity, but not before. Irrespective of the exact age at which sensory input begins to influence motor activity, there is at least a period in early development when activity does not depend on sensory stimulation in any immediate sense (Hamburger, 1973). However, many of the earlier researchers considered the avian embryo to be much more insulated from the environment than we now know it to be (for further discussion of this point see Vince, 1973).

At the time of the earlier research there was also support for the hypothesis that self-stimulation is important for the development of behaviour patterns displayed by the embryo and also the hatched chick (Kuo, 1932a,b; Lehrman, 1953; Schneirla, 1965). Kuo (1932b), for instance, believed that the heart-beat of the embryo stimulated rhythmic, passive lifting of its head and that this eventually gave rise to the pecking response. Furthermore, contractions of the amnion and yolk sac were considered to stimulate embryonic motility. Contractions of the amnion begin on day E4 and increase in frequency and amplitude to day E9 (Gottlieb and Kuo, 1965). Kuo believed that myogenic contractions of the amnion cause passive movement of the embryo, and that these serve not only to prevent adhesions forming between the embryo and the surrounding membranes, but also develop into active movements of the embryo. Contractions of the amnion, however, have not been observed to modify the rhythmical pattern of activity of the embryo (Hamburger *et al.*, 1965; Oppenheim, 1966). There are clearly active movements generated by the embryo itself, and these begin as early as day E3 or E4. Ultimately, however, substantiation of the self-stimulation hypothesis requires knowledge of the stage of development at which the tactile sensory system becomes functional and able to provide perceptual input to the developing nervous system, rather than the stage at which it is linked to motor responses in the embryo. Furthermore, it would be necessary to demonstrate that this sensory input influences later behaviour, after hatching and possibly even in adulthood. Despite the fact that research in this area is no longer fashionable, this deserves further investigation.

Hatching

Hatching is a remarkable event involving important embryological as well as general biological problems. These have been clearly delineated by Oppenheim (1973). The following is a discussion of only the behavioural events of hatching.

The embryo prepares itself for the hatching process, or climax, by orienting into the correct position and performing specialized movements. Prehatching behaviour begins on day E16 or E17 with the appearance of new behaviour patterns (Hamburger and Oppenheim, 1972b; Oppenheim, 1967). Before this stage the movements of the embryo are primarily random, jerky and convulsive, but at prehatching they become smooth and coordinated. That is, they become more organized and stereotyped, eventually leading to the coordinated rotary movements necessary for hatching.

On day E18 the chick orients into the tucking position, in which the head lifts up from being buried in the yolk sac and becomes turned to the left side of the body. This is the position adopted by at least 95% of embryos, but the remaining small number orients in the opposite direction (Oppenheim, 1973). In the usual position, the beak is positioned against the membrane of the air sac, the left eye is occluded by the body and the right eye is placed next to the air sac membrane (Hamburger and Oppenheim, 1967). The right wing partially covers the right side of the head. Vigorous thrusts of the head and beak occur in the direction of the air sac membrane. Towards the end of day E18 the embryo has attained the hatching position.

On either day E19 or E20 the beak penetrates the membrane of the air sac, as a result of a vigorous head movement (Oppenheim, 1970; Fig. 1.3). Consequently, breathing can begin and vocalizations commence (Oppenheim, 1972b). In addition, clearing of the nares makes olfactory discrimination possible (see p. 37). The frequency of beak clapping rises during day E19, and movements of the head are more common than those of the body. In fact, vigorous whole body movements (designated as type III movements by Hamburger and Oppenheim, 1967) occur rarely between tucking and climax. Nevertheless, when they do occur, these movements are responsible for changes in position of the embryo.

Eventually one of the vigorous head movements causes cracking of the eggshell in one spot, known as pipping. This can occur on day E19 or E20, some 24 hours before hatching. In eggs positioned normally, through the action of gravity, either in natural incubation conditions or in an incubator the shell is pipped on the uppermost side (Oppenheim, 1973), which apparently assists respiration (Kovach, 1968). The hole is not usually enlarged until hatching, probably because head movements decrease after pipping and until hatching. In fact, the period from pipping to hatching is one of relative quiescence, also apparent in brain cell inactivity (see pp. 33–36).

The onset of hatching is signalled by increased activity. A series of back

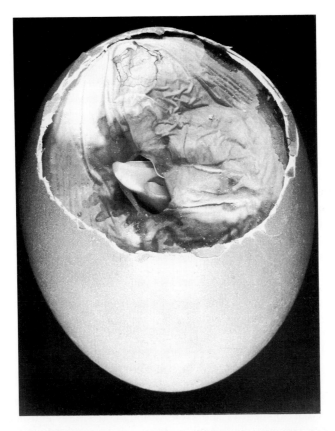

Fig. 1.3. The position of the chick embryo at day E20 of incubation. The embryo is in the hatching position and the beak has penetrated the membrane into the air sac.

thrusts of the beak, with its egg tooth, against the shell enlarges the hole originally made at pipping. These continue as the body rotates in the egg, aided by stepping movements of the feet against the eggshell (Hamburger and Oppenheim, 1967). The direction of rotation is anticlockwise when viewed from the air sac end of the egg. During the hatching process the respiration rate increases markedly (Oppenheim, 1972b). The coordinated movements that occur during climax are controlled by the forebrain (Oppenheim, 1972a), and they are at least in part due to sensory feedback produced by the embryo's position in the confined space of the egg. If the embryo is removed from the egg at this stage, the typical movements cease (Corner and Bakhuis, 1969). Oppenheim (1973), however, tended to be less convinced of the result since he found little disturbance of these movements in embryos partially pulled out of their eggs. Other factors, either neural or hormonal, may well be involved in initiating and maintaining hatching behaviour.

Each time the chick displays the struggling movements characteristic of climax, the eyes are open, vocalizations are likely to be made and the EEG indicates an awake state (Bakhuis and Bour, 1980). Paradoxical sleep patterns begin to appear three days before hatching, and sleep–wake cycles appear to be present by hatching (Corner *et al.*, 1973). Earlier researchers had suggested that the rhythmical movements associated with hatching might occur during episodes of paradoxical sleep (Peters *et al.*, 1965), but the subsequent study by Bakhuis and Bour (1980) demonstrated that these movements occur during the awake state. It might, therefore, be during the struggling movements when the chick embryo is awake that it attends to environmental inputs, and possibly learns. Tschanz (1968) has suggested that it is at this time that guillemot embryos learn the specific calls of their parent. When the embryo moves and vocalizes, the hen responds by rising, rolling the egg and calling. Thus, an awake state of the embryo is synchronized with parental calling, and so learning may be maximized. This may be the mechanism by which the guillemot embryo learns its own parent's calls and not those of neighbouring hens nesting close by on the ledge. Tuculescu and Griswold (1983) have investigated embryo–hen vocal interactions in domestic chicks, and it appears that a similar situation occurs in this species (Chapter 2).

Development of the Sensory Systems

The various sensory systems become functional at different times during embryonic development (Table 1.1). Although structures and neural pathways which comprise the visual, auditory, olfactory, tactile and vestibular systems all begin to differentiate on day E2, each develops at its own rate so that sensitivity to stimuli in the various modalities appears according to a sequence. In general terms, it begins with tactile sensitivity on day E6, followed by vestibular sensitivity on day E8, auditory sensitivity on day E12 to E14, visual responsiveness beginning on day E18 and improving over the last days of incubation, and olfactory responsiveness on day E20. The sense of taste presumably develops rather early, possibly by day E12 when the embryo begins to imbibe the amniotic fluid, although this is not known. It is also possible that fluid-borne chemicals reach the olfactory epithelium via the back of the palate and stimulate it prior to day E20.

The orderly developmental sequence of the different sensory modalities appears to be important for the integration of information received from the various sensory systems after hatching. For example, the fact that the auditory system of the embryo matures before the visual system allows auditory inputs to be received by the developing brain, and auditory memories to be formed, without competition from visual inputs. Changing this order of development and stimulation alters intermodal sensory functioning and social behaviour after hatching (Lickliter, 1990; Banker and Lickliter, 1993). Of course, proprioceptive inputs may well interact with the auditory inputs, as these pathways have developed earlier, but this potential interaction has not yet been studied. More discussion of this topic appears in Chapter 2.

The development of each of the sensory systems will be considered separately at first (in this chapter), followed by discussion of the interactions between these systems during embryonic development (Chapter 2). The reader is advised to refer to Table 1.1 when reading the following sections.

Tactile, proprioceptive and vestibular sensitivity

Sensory fibres from the facial ganglion begin to enter the brain on day E2, but motor responses to tactile stimuli are not yet present. Some of the somatosensory and somatomotor regions of the brain have developed by day E5: these are the nucleus basalis and archistriatum ventrale, respectively (Tsai *et al.*, 1981). Neurons destined for the hyperstriatum accessorium (a region which processes both somatosensory and visual information) form on day E5 and continue to differentiate for the next five days (Tsai *et al.*, 1981). Tactile sensitivity can be demonstrated on day E6 to E7 by touching the oral region of the embryo with a fine hair, in response to which the embryo shows withdrawal movements (Hamburger and Balaban, 1963). Thus, by this age the reflex arc has been established. By day E10 local proprioceptive muscle reflexes occur (research cited by Gottlieb, 1968). The presence of these reflexes is important for the correct development of the bones and joints to occur.

Tactile conditioning has been studied in chick embryos during the later stages of incubation (research by a number of authors cited by Gottlieb, 1968). At the final stages of its development the embryo is able to habituate to cutaneous stimulation, and the rate of habituation improves significantly from day E17 to E20. Over a similar period, there is improved ability for conditioning to cutaneous stimulation associated with an electric shock. The latter is, of course, an artificial paradigm not normally encountered by a developing embryo. In fact, the embryo is protected from most potentially punishing environmental inputs, except perhaps temperature drop, or changes in humidity.

On the other hand, cutaneous stimulation of the embryo may be provided by movement of the amnion or other membranes, or by self-stimulation when the embryo moves. It may also occur as a consequence of the hen's turning the eggs in the nest, a behaviour which she performs at regular intervals throughout the incubation period (Gottlieb, 1968).

Turning of the eggs would also provide stimulation of the vestibular system. The turning temporarily displaces the embryo and then it moves back to the uppermost side of the egg. Neurogenesis in the vestibular ganglion begins as early as day E2, but vestibular sensitivity has not been demonstrated until day E8. On day E8, rotation of the egg on a disc generates nystagmus of the head (head oscillations), followed by a form of head-shaking after the rotation has ceased (studies by Visintini and Levi-Montalcini, 1939, cited by Gottlieb, 1968, and Freeman and Vince, 1974). By contrast, Kuo (1932a) failed to find active movement in response to rotation until a much later stage when the embryo's beak had penetrated the air sac membrane. In the embryo the vestibular system appears to be important for inhibiting excessive activity and for controlling the rotary head movements used in hatching (Gottlieb, 1968).

The stage of development at which proprioceptive sensation first appears is not definitely known, but it may be at about day E10 (Freeman and Vince, 1974). In general, therefore, all of these sensory systems (tactile, proprioceptive and vestibular) become functional rather early in embryonic development, and so provide the first sensory inputs controlling behaviour.

Auditory sensitivity

Development of neuronal connections in the auditory system

The auditory system begins to differentiate on day E2 and continues to develop throughout the incubation period. By day E12 the embryo can already respond to auditory stimulation by changing its behaviour, and prior to hatching the embryo can perceive and imprint on the hen's calls (Grier *et al.*, 1967). Thus, early development of the auditory system seems critical for the survival of the newly hatched chick.

On day E2 the inner ear begins to develop, followed on day E3 by the eighth nerve beginning to grow into the brain (Vazquez-Nin and Sotelo, 1968). The latter can be seen entering the first auditory relay nucleus, the nucleus magnocellularis, between days E6 and E7 (Jackson *et al.*, 1982; Rubel *et al.*, 1976). Neurogenesis in the cochlear ganglion continues until day E7 or E8, which is consistent with the time of entry of the axons from this ganglion (which form the eighth nerve) into the auditory nuclei. The sensory hair cells in the cochlea appear on day E8 (Rebollo and Casas, 1964, cited by Jackson and Rubel, 1978), and neural connections to these hair cells take on a mature appearance by day E13.

The primary auditory region of the forebrain is known as Field L (Chapter 4, Fig. 4.3) and this region begins to differentiate on day E8 (Tsai *et al.*, 1981). Field L receives inputs from the relay auditory nuclei. Once it is fully developed it processes higher order aspects of auditory information (Heil *et al.*, 1992).

Following the period of neurogenesis, and growth of the eighth nerve into the auditory nuclei, there is a period of cell death that occurs between days E11 and E13 in the nucleus magnocellularis and another auditory nucleus to which it projects, the nucleus laminaris (Jackson *et al.*, 1982). In fact, the nucleus magnocellularis loses 18% of its neurons and the nucleus laminaris loses as much as 84% of its neurons (Rubel *et al.*, 1976). Cell death is a normal aspect of the development of sensory systems (Oppenheim *et al.*, 1992). It is part of the fine tuning of neuronal connections, often occurring in response to sensory input. Indeed, the first acoustically evoked potentials can be recorded from the brain on day E11 (Saunders *et al.*, 1973), but only to intense auditory stimuli. Thus, whether the cell death in the auditory nuclei would occur in response to sensory input in the natural environment is unknown. No measurement has been made of the sound intensity within the egg of the hen's calls, although it is known, for hens and other species, that the maternal bird vocalizes when she is turning the eggs and therefore when she is in close proximity to the eggs (Tschanz, 1968; Guyomarc'h, 1972). Consequently, the embryo may be exposed to species-specific high intensity sounds. These sounds might be perceived by the embryo at somewhere between day E11 and E14 depending on their intensity and other characteristics.

The eighth nerve is certainly able to transmit information to the brain by day E11, and at this stage postsynaptic responses can first be obtained from the nucleus magnocellularis (Jackson *et al.*, 1982). One day later, on day E12, they can be recorded from the nucleus laminaris (Jackson *et al.*, 1982). Within the nucleus magnocellularis the responses to stimulation of the eighth nerve can first be detected in the anteromedial region, a region which responds to high frequencies (Rubel and Parks, 1975), and only later can they be detected in other regions of the nucleus. By day E13 all areas of the nucleus have been innervated (Jackson *et al.*, 1982).

Tonotopic organization and the pattern of development

It should be noted here that the nucleus magnocellularis and the nucleus laminaris are tonotopically organized (Rubel and Parks, 1975). That is, the neurons within the nucleus are placed in an orderly arrangement with respect to their preferred tones. In fact, the response preferences are arranged in isofrequency columns within the nucleus, and these are arranged tonotopically over the nucleus. Electrophysiological recordings from units in the nucleus magnocellularis can be obtained from ipsilateral, monoaural stimulation. The nucleus laminaris receives binaural inputs that are also tonotopically organized.

These data have been obtained from 5- to 15-day-old chicks. It is not known when the tonotopic organization first develops. Field L of the forebrain is also tonotopically organized (Heil *et al.*, 1992). It is not known whether this organization develops in the embryo, but this would appear to be likely at least in a basic form that is modified or refined with later experience.

The appearance of auditory responses

The gradient of growth of neural inputs to the nucleus magnocellularis matches the basal-to-apical gradient of development of the basilar papilla within the ear (that is, from the region which responds to high frequencies to the region which responds to low frequencies). Therefore, it might be expected that behavioural responses of the embryo to high frequency sounds would precede those given to low frequency sounds, but this may not be so. There are conflicting results. On day E12 the embryo moves in response to low frequency auditory stimulation only (Gottlieb, 1968; Rubel, 1978). However, Jackson and Rubel (1978) played 400, 700 and 1400 Hz tones to embryos on day E14 and found that, whereas both the 700 and 1400 Hz tones caused a decrease in their activity, there was no response to the low frequency tone of 400 Hz. By day E16 activity increases in response to the 700 and 1400 Hz tones, and not until day E17 does the 400 Hz tone produce a response of increased activity. These results do, in fact, suggest that behavioural responsiveness develops first to higher frequencies. For duck embryos, however, Gottlieb (1979) has found that responsiveness develops first to low frequency sounds and later to high frequency sounds.

Yet other contradictory results have been obtained for chicks after hatching. At this stage of development responsiveness to low frequency sounds appears to develop before that of high frequency sounds (Gray and Rubel, 1985). In all of these procedures, however, one must consider the way in which responsiveness is measured. Gray and Rubel (1985) for example, assessed the auditory thresholds of response to sounds of various frequencies by noting whether the isolated chick ceased peeping when the tone was played. Therefore, not only must the chick hear the tone above its own peeping, but the tone must be behaviourally important to the chick. Similar variables would also apply to testing the auditory responsiveness of chick embryos.

It is, of course, also possible that the basal portion of the papilla of the young embryo could respond to low frequency sounds and that this may change with age, as suggested by Jackson *et al.* (1982). This aspect of development deserves further research. Furthermore, Vince *et al.* (1976) have found that click sounds alter the motility of the embryo when presented on day E15, although it is uncertain whether this stimulus is vibratory rather than acoustic.

Auditory thresholds measured electrophysiologically at the basilar papilla and round window of the ear show developing maturation from day E15 until the first day posthatching (Rebillard and Rubel, 1981). Here, paradoxically, the electrophysiological responses mature first to low frequencies and later to high frequencies. Similarly, evoked-responses recorded from the cochlear nuclei occur first to low frequency sounds, the maximum sensitivity shifting from about 400 Hz on day E13 to 800 Hz on day E18 or E19 (Saunders *et al.*, 1973). Clearly, there is a need for more detailed study of the relationship between behavioural maturation of responses to high versus low frequencies and the electrophysiological and anatomical time-courses of development.

The threshold of auditory detection is considerably reduced on day E19, when the middle ear is cleared of fluids (Saunders *et al.*, 1973). Also on day E19, myelination of the eighth nerve, which commenced on day E11, has been completed (cited by Dmitrieva and Gottlieb, 1992), and consequently the conduction velocity of this nerve increases (Jackson *et al.*, 1982).

Cochlear microphonic potentials have been recorded from day E13 embryos by Vanzulli and Garcia-Austt (1963), and by day E11 by Saunders *et al.* (1973). Histological observations indicate that the cochlea is mature by about day E15 (Saunders *et al.*, 1973). As already discussed, the first behavioural responses to auditory stimuli have been obtained on day E12, and these mature throughout the incubation period. As early as day E16, the embryo shows some capacity for conditioning to auditory stimuli, but this capacity improves at around day E19 or E20 (Sedláček, 1964). On day E19/20 the embryo's capacity for auditory conditioning increases, and this is the stage at which the embryo learns to identify the hen's calls (Chapter 2).

Thus, auditory sensitivity develops early in the second half of incubation and continues to develop through the remainder of the incubation period. Before hatching, the auditory system has matured sufficiently for the chick to the be able to learn auditory preferences.

Visual sensitivity

Chickens have well-developed vision with spectral sensitivities ranging from the infrared to the ultraviolet regions of the spectrum, and much of this visual capacity is known to develop before hatching.

Neurons of the optic vesicles differentiate as early as day E2 (Prada *et al.*, 1991), while axons from the retinal ganglion cells begin to leave the retina, growing towards the brain (Kahn, 1973; Rager, 1980). These axons reach the optic chiasma by day E4 and advance further along the optic tract to reach the contralateral optic tectum by day E6 (Thanos and Bonhoeffer, 1983). Pigmentation of the eyes begins on day E3 (Romanoff, 1960) and there is a tendency for the right eye to pigment first (Harrison, 1951). This detail will become relevant to the discussion in Chapter 2.

Also at this stage of development, exposure of the embryo to light produces increased activity of the embryo. Bursian (1964, cited in Gottlieb, 1968) found that the frequency and duration of activity cycles were increased by light exposure (10–20 s of a 500 W white photo lamp, which provides intense light not likely to be present during normal conditions of incubation). Further experiments using embryos on day E4 revealed wavelength specificity for this response: exposure to green light was found to increase the duration of the activity cycle, whereas red and blue had no effect (Bursian, 1965). It has been assumed that the response is mediated non-visually, possibly via photoreceptors in the skin, muscle or brain (Gottlieb, 1968). Although it is conceivable that even by day E3 some axons from the retinal ganglion cells have reached the brain, no synapses between ganglion cell axons and cells within the optic tectum have been detected this early in development and, moreover, the photoreceptors in the retina do not differentiate until the later stages of incubation. Therefore, it is not likely that these projections could mediate visual input from the eyes to the brain to cause the increased motility. Alternatively, a role for light-sensitive cells in the pineal gland might be considered in connection to the increased motility because the pineal has formed by day E3 or E4 (Romanoff, 1960). Alternatively, the influence of light may be mediated via photoreceptor cells within the developing hypothalamus: photoreceptor cells are known to be present in the hypothalamus of Japanese quail, where they act to control gonadotropin-releasing hormone in response to changing day length (Perera and Follett, 1992).

The increased activity in response to white light continues until day E6, and then from day E6 to E8 activity decreases in response to exposure to white and green light, whereas it still increases in response to intense red light stimulation (Bursian, 1965). Blue light remains ineffective during this period. Sensitivity to blue light appears to develop much later, by day E18, when the embryo can discriminate between red and blue light (Sedláček, 1972). However, Garcia-Austt and Patetta-Queirolo (1961) found that blue light does not evoke an electroretinogram until day E19.

Development of the retina

The first cells to differentiate in the embryonic retina are the ganglion cells (day E2). The amacrine cells, which integrate information horizontally across the retina, are generated from day E3 to E9 (Kahn, 1974; Spencer and Robinson, 1989) and the subpopulation of dopaminergic amacrine cells forms between day E3 and E7 (Gardino *et al.*, 1993). The neurotransmitters GABA and acetylcholine are present by day E6 (Seiler and Sarhan, 1983; Araki *et al.*, 1982). The cholinergic cells are the first ones to differentiate in the retina, and on day E10 they have fully developed (Spira *et al.*, 1987). After this, synaptogenesis begins within the retina (Fisher, 1983). The latter, together with an increased dendritic branching of neurons, causes a thickening of the inner plexiform layer at day E10 (Coulombre, 1955). The photoreceptors differentiate between days E10 and E12 (Romanoff, 1960). The bipolar cells of the retina are the last neural cells to form, between days E11 and E13, (Prada *et al.*, 1991), indicating that the retina cannot assume its full integrative capacity before day E13 or perhaps somewhat later. The dendritic plexus continues to develop from day E13 to E16 (Gardino *et al.*, 1993).

The receptors for the amino acid neurotransmitters, used by many neurons in the retina, develop much later than those for acetylcholine. Aspartate receptors increase at day E14 and NMDA (*N*-methyl-D-aspartate)-type receptors for glutamate do not increase until day E18 (Somohano *et al.*, 1988), which coincides with the first measurable electroretinogram (see p. 29).

Glial cells form in the inner retina during days E12 to E17, and on day E14 the oligodendrocytes begin to wrap around the nerve axons to form the myelin sheaths (Nakazawa *et al.*, 1993). Myelination of the fibres in the optic nerve begins on day E15 (Rager, 1980), and continues until about one week after hatching (Nakazawa *et al.*, 1993).

Development of the optic tectum

The time at which the embryo's response to white light changes from increased to decreased activity (day E6) coincides with an important step in the development of the visual projections. The first axons from the retinal ganglion cells arrive at the optic tectum on day E6 and begin to advance over its surface, to cover it by day E13 (Rager and von Oeynhausen, 1979). Branching of these neurons within the optic tectum begins to occur on day E9. On day E9 glial cells are first detectable in the optic tectum and they continue to differentiate through the incubation period (Linser and Perkins, 1987). The first mature synapses are found in the optic tectum on day E8 according to McGraw and McLaughlin (1980) and by day E11 according to a number of other researchers, including Rager (1976) and Panzica and Viglietti-Panzica (1981).

During days E10 to E11 there is a marked increase in the number of arborizing axons in the optic tectum and the size of the axonal trees and their amount of branching continues to increase at least until day E18 (Thanos and Bonhoeffer, 1987). These projections are initially imprecise and will later be refined, with elimination of up to 40% of the ganglion cell axons (Thanos and Bonhoeffer, 1983). On day E12 the cells in the optic tectum start growing and forming functional connections (Miyake and Morino, 1992). This has been discussed in detail by Mey and Thanos (1993).

The reader is reminded that throughout the period from day E6 to E13 when the optic tectum is differentiating, the retina is also continuing to differentiate. Within the optic tectum rapid growth of cells and synaptic formation occurs on day E12. Myelination of the optic fibres is complete by day E15 (Schifferli, 1948), and low amplitude electrical potentials can be recorded from the optic tectum on day E14 or E15 (Sedláček, 1967). On day E15 also the pupillary reflex is neurally mediated (Heaton, 1976), and ocular movements occur in response to visual stimulation (Sedláček, 1967). Axonal arborization in the optic tectum is well advanced by day E16, and at this time the retinotectal system is structurally mature. Nevertheless, electrical activity cannot be recorded from the retina until day E17. As might be expected, visually evoked potentials can be recorded from the optic tectum at the same stage of development (Peters *et al.*, 1958). A typical electroretinogram is, however, not obtainable until day E18 (Patetta-Queriolo and Garcia-Austt, 1956), and does not assume a postnatal-like form until day E19 (Garcia-Austt and Patetta-Queriolo, 1961). By day E19 light stimulation evokes increased beak clapping (Oppenheim, 1968) and increased opening and closing of the eyelid (Kuo, 1932a).

The isthmo-optic pathway

The isthmo-optic nucleus, which will send efferent nerves to the retina, develops according to its own programme. This nucleus is situated in the midbrain not far from the optic tectum from which it receives input. Thus, it forms part of a feedback loop to the retina (Fig. 1.4). One of its functions involves increasing visual sensitivity when the chick turns to look into areas of low light intensity (Rogers and Miles, 1972). For example, if the chick turns to look for food grain, or more particularly a small, moving beetle, in a dark area such as under a leaf, its vision will be made more sensitive by the action of the isthmo-optic system. In fact, the isthmo-optic nucleus is well-developed

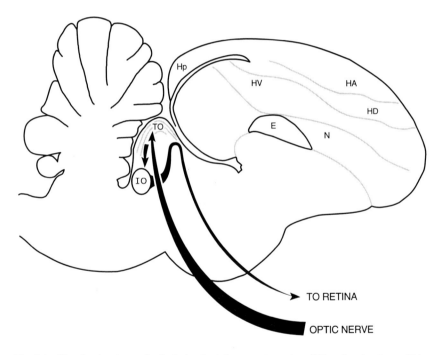

Fig. 1.4. The visual pathway that includes the isthmo-optic nucleus (IO) and projections which provide feedback to the retina. The pathway begins with projections from the ganglion cells in the retina to the contralateral optic tectum (TO). From TO there are projections to IO and from IO projections travel in the optic nerve back to the contralateral eye. Although not illustrated, in the optic chiasma there is almost total crossing over of both the retinal ganglion cell projections from each eye and the projections from each IO to the retina of each eye. Other abbreviations: Hp, hippocampus; HA, hyperstriatum accessorium; HD, hyperstriatum dorsale; HV, hyperstriatum ventrale; E, ectostriatum; N, neostriatum.

in ground-feeding birds such as the chick, quail and pigeon (Uchiyama, 1989). Even though the efferent fibres from the isthmo-optic nucleus make up only 0.4% of the total fibres in the optic nerve, in ground-feeding birds this proportion is tenfold greater than in raptors (Weidner *et al.*, 1987). This reinforces the suggestion that the isthmo-optic visual system is involved in food detection in the chicken. The isthmo-optic projections enhance the sensitivity of the retina at the expense of sensitivity to contrast (Uchiyama, 1989). The net effect is to speed the rate of detection of novel objects, be they food objects or predators (Rogers and Miles, 1972). The projections from the isthmo-optic nucleus go to the ventral retina of the contralateral eye (Fritzsch *et al.*, 1990) and, via amacrine cells, connections are made to the dorsal retina. Thus, it is also possible that the isthmo-optic nucleus plays a role in switching attention between the upper and lower visual fields (Catsicas *et al.*, 1987). This role may be important in a species, like the chicken, with large eyes that are not moved to any great extent.

The neurons which will form the isthmo-optic nucleus are formed between days E5 and E7, and they migrate to their correct place in the brain from day E7 to about day E9 (Clarke, 1992). Axons from the nucleus begin to grow towards the retina on day E7 and reach it on days E9 and E10 (Fritzsch *et al.*, 1990). From day E12 on, synapse formation occurs with the amacrine cells in the retina, and is completed by day E18 (Clarke, 1992).

The afferent projections to the isthmo-optic nucleus from the tectum do not form until about day E13 or E14, and their arrival at the nucleus triggers programmed cell death of more than half the axons from the isthmo-optic nucleus to the retina. Apparently aberrant axons projecting to the 'wrong' part of the retina are eliminated (Clarke, 1992). This process is completed on day E16. Hence, the isthmo-optic, feedback loop of the visual system is largely mature by day E16, although the nucleus undergoes further differentiation after hatching.

Development of the visual projections to the forebrain

The development of the visual projections from the optic tectum to the forebrain regions has been less studied. Also, some of the retinal ganglion cells project to the rostral region of the thalamus, to nuclei known collectively as the nuclei opticus principalis (OPT), and there has, to my knowledge, been no investigation of the development of these projections.

There are two main sets of projections to visual regions of the forebrain (Fig. 1.5 and Engelage and Bischof, 1993). One from the optic tectum projects first to the ipsilateral nucleus rotundus (ROT), and from there to the ectostriatal region of the forebrain. This is known as the tectofugal visual system. The other set of projections leaves the opticus principalis (OPT) in the thalamus and goes to the hyperstriatal region of the forebrain. This is known as the thalamofugal visual system. There are both ipsilateral and contralateral

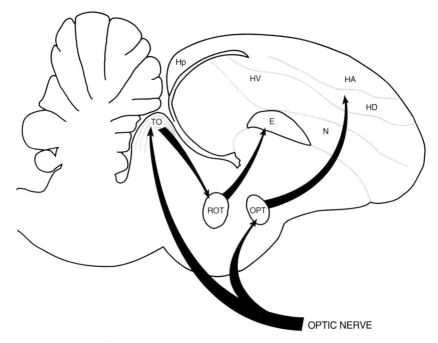

Fig. 1.5. The thalamofugal and tectofugal visual pathways. The thalamofugal pathway consists of projections from the retina to a collection of nuclei in the thalamus, known as the nuclei opticus principalis (OPT), and from here there are projections to the hyperstriatum dorsale (HD). The tectofugal system consists of retinal projections to the optic tectum (TO), from here another set of projections goes to the nucleus rotundus (ROT) and a further set go to the ectostriatum (E). Abbreviations are as in Fig. 1.4.

projections from the OPT on each side of the thalamus to the hyperstriata on each side of the forebrain.

Information obtained by placing lesions in adult pigeons shows that the two visual systems control somewhat different visually guided behaviours. The tectofugal pathway is clearly involved with colour vision, visual acuity, detection of movement and pattern discrimination, whereas the thalamofugal system may also be involved in pattern discrimination and detection of movement, but it has no role in colour vision or visual acuity (Pritz *et al.*, 1970; Hodos, 1976; Wilson, 1980; Macko and Hodos, 1984; Güntürkün *et al.*, 1989). The function of the thalamofugal system remains equivocal at present. Although Shimizu and Hodos (1989) reported data showing that it might be important for reversal learning of visual stimuli, Chaves *et al.* (1993) have disputed this claim and suggested that their results reflected functioning of the tectofugal system instead. There is evidence that the thalamofugal visual system may be used for acquisition and reversal of spatial discrimination tasks with long, but not short, inter-trial intervals (Macphail and Reilly,

1989). The thalamofugal system may also function to detect stimuli in the lateral field of vision and direct eye and head movements so that a stimulus can be viewed in the frontal, binocular field for more detailed inspection using the tectofugal system (Maxwell and Granda, 1979; Güntürkün *et al.*, 1989), although there is not a great deal of evidence in support of this hypothesis.

The development of the thalamofugal visual projections involves growth of ipsilateral projections and contralateral projections from the thalamus. The contralateral projections cross the midline of the brain at the thalamic level in the supra-optic decussation (SOD). The number of axonal fibres in the SOD increases from day E12 to E19 and then declines from day E19 and throughout the first week of posthatching life (Rogers and Ehrlich, 1983; and Chapter 4). Thus, development of the thalamofugal visual system continues after hatching. Indeed, after hatching it develops an asymmetry in the number of contralateral projections which cross in the SOD, and it does so in response to exposure of the embryo to light during the last two or three days before hatching (Chapter 2).

The tectofugal visual system may become functional before the thalamofugal visual system. The ectostriatum in the forebrain has formed by as early as day E5 (Tsai *et al.*, 1981), although it must continue to differentiate throughout the incubation period. The metabolic activity of the ectostriatal region is relatively high on day E19 and also on the first day after hatching, as indicated by the uptake of radioactively labelled 2-deoxyglucose (2-DG) by neurons (Rogers and Bell, 1989, 1994; Fig. 1.6). This sugar is taken up by neurons but, as it cannot be metabolized by them, its accumulation indicates neuronal activity, either as a reflection of firing activity or growth processes. It is known that the metabolic activity levels in the ectostriatum of one-day-old chicks are determined by visual stimulation, as stimulation of one eye only causes more uptake of 2-DG in the contralateral ectostriatum (Rogers and Bell, 1989). In contrast to this early visual responsiveness in the tectofugal system, development of the thalamofugal system seems to be delayed. Regions of the hyperstriatum which receive the thalamofugal visual projections remain relatively inactive until the end of the first week posthatching.

The projections from the OPT have their end terminals primarily in a region of the hyperstriatum known as the hyperstriatum dorsale (HD) (Karten *et al.*, 1973), and an adjacent region known as the nucleus intercalatus of the hyperstriatum accessorium (IHA) (Dubbeldam, 1991). From the HD there are projections to the hyperstriatum accessorium (HA) (Fig. 1.6). These projections must relay visual information to the HA because application of electrical stimuli to the optic papilla evokes responses in the HA (Boxer and Stanford, 1984) and furthermore visual stimuli directly evoke responses in the HA (Wilson, 1980), as also do somatosensory inputs (Deng and Wang, 1992).

On day E20 just after the embryo has pipped the egg shell, on the first day of posthatching life and until the end of the first week posthatching, the HA is relatively inactive, again as determined by measuring the uptake of

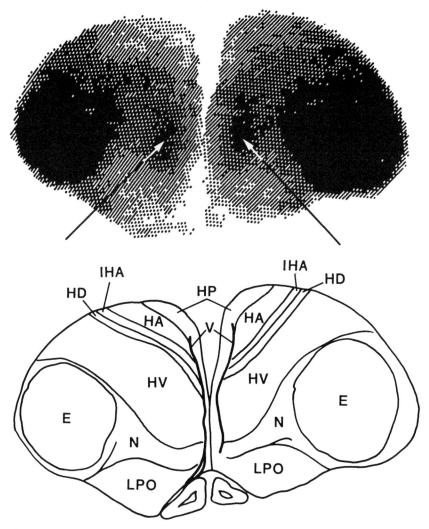

Fig. 1.6. A transverse section taken at approximately 3 mm from the rostral pole of the forebrain of a one-day-old chick. The chick was injected with 2-deoxyl[^{14}C]glucose (2-DG) and allowed 30 min of light exposure in the home cage with food and another chick present to provide ample visual stimulation. Note the high uptake of 2-DG, indicating high metabolic activity, in the ectostriatum (E) and in the intermediate medial region of the hyperstriatum ventrale (IMHV) (arrows). There may be a link between activity in the tectofugal visual system and in IMHV. The latter is the site of memory formation in early life, the high activity thus indicating that the chick may be imprinting. By contrast, low activity occurs in the regions of the forebrain which receive direct input from the thalamofugal visual system (i.e. hyperstriatum dorsale, HD; nucleus intercalatus hyperstriati accessorii, IHA) as well as that which receives secondary input from this visual system (i.e. the hyperstriatum accessorium, HA). The uptake of 2-DG by the hippocampus (HP) is also low. The other labels are neostriatum (N), lobus parolfactorium (LPO), hyperstriatum ventrale (HV), forebrain ventricle (V). From Rogers and Bell (1994), with kind permission from Elsevier Science Ltd.

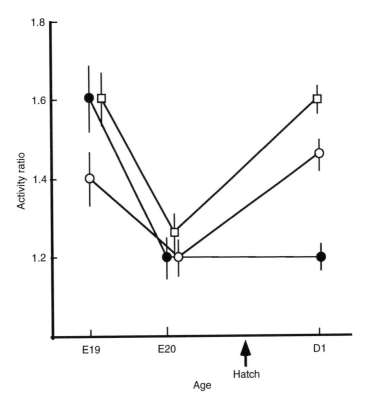

Fig. 1.7. Metabolic activity in three regions of the chick forebrain measured by uptake of 2-deoxy[^{14}C]glucose injected 30 min before death at either day E19 or E20 of embryonic life or day 1 (D1) posthatching. Sections were made of the forebrain, and examined by a computerized imaging process. Optical density measurements were taken for the hyperstriatum accessorium (HA●), hyperstriatum dorsale (HD) ○ and the intermediate medial hyperstriatum ventrale (IMHV) □. The values were expressed as ratios to a region of low metabolic activity (hippocampus) on the same section. Note the low levels of metabolic activity in all three regions on day E20, just after pipping has occurred. By day 1 posthatching, the activity levels of HD and IMHV have risen, whereas activity remains low in HA. The figure is modified from data presented in Rogers and Bell (1989, 1994) and Bell and Rogers (1992).

radioactive 2-deoxyglucose by neurons (Bell and Rogers, 1992; Rogers and Bell, 1994). Prior to this metabolically quiescent period in the HA, on day E19 the level of metabolic activity in the HA is considerably higher (Fig. 1.7), although at this stage it is not visually driven and so the uptake of 2-deoxyglucose at this stage may simply reflect growth processes (Rogers and Bell, 1994). Hence, metabolic activity in the HA is suppressed prior to hatching and throughout the posthatching period when early learning is occurring. HA metabolic activity rises again towards the end of the first week of life. Of course, it is possible that low activity in the HA means that there is low background activity and that only specific information is being processed in the HA (i.e. at levels of activity too low to be detected by 2-deoxyglucose uptake), but clearly visual driving of the HA does not fully mature until the end of the first week of posthatching life.

The region of the brain in which imprinting memories are stored, the intermediate medial hyperstriatum ventrale (IMHV) (Fig. 1.6; Chapter 4 covers this in more detail), receives visual inputs from both the thalamofugal system via the HA, and the tectofugal system via the ectostriatum. Therefore, since HA activity is suppressed after hatching, it would seem that imprinting might rely on visual information from the tectofugal system, and not the thalamofugal system. Nevertheless, metabolic activity in the HD, which receives the end terminals of the projections from the OPT, has been measured as relatively high on both day E19 and on day 1 posthatching (Fig. 1.7). The information being processed in the HD is apparently not being relayed to the HA, and thus on to the IMHV.

Of course, limited but specific information, not causing sufficient neural activity to be detected by 2-DG uptake above the background level, might be relayed to the HA and on to the IMHV at this stage. There is simply not activation of HA neurons in general. On day E20, however, the metabolic activity in the HD is significantly lower than either before this stage or after hatching (Fig. 1.7), which signifies a temporary suppression of activity prior to (and probably during) hatching. The metabolic activity of the IMHV follows the same pattern as the HD (Fig. 1.7). Overall, therefore, there is a decrease in metabolic activity in the HA, HD and IMHV on day E20, prior to hatching. At the same time, activity in the ectostriatum remains relatively high (Rogers and Bell, 1989, 1994). Consequently, the tectofugal visual system is likely to have dominance during the period just before and immediately after hatching. It should be noted, however, that these data are for embryos exposed to light before hatching. The onset of the suppression of HA activity in particular may differ in dark-incubated embryos. This is discussed further in Chapter 4.

Late embryonic behaviour in response to visual stimuli might therefore be dominated by the tectofugal system, which, in pigeons at least, has been shown to be important in pattern discrimination (Hodos and Karten, 1970), colour vision (Hodos, 1969), visual acuity (Hodos *et al.*, 1984) and size-discrimination (Hodos *et al.*, 1986). On day E18 the embryo can discriminate red from blue, and by day E21 it is able to discriminate red, blue and orange wavelengths (Sedláček, 1972). This ability almost certainly reflects functioning of the tectofugal system. Similarly, the light-evoked potentials which can be first recorded from the chicken forebrain on day E19 (Sedláček, 1967) must be largely, if not exclusively, generated by the tectofugal system.

In summary, the visual system is differentiating throughout embryonic development and even up to three weeks after hatching (Chapter 4), but various responses to light stimulation can be detected throughout this period. From day E3 to day E8 changes in motility of the embryo result from light stimulation. Discrimination of some wavelengths becomes possible by day E18. At precise times throughout incubation various aspects of the visual system assume functional maturity, but forebrain involvement in visual processing appears to be delayed until day E19 or day E20, and then it may occur in the ectostriatal system only. Apparently, not until much later will the thalamofugal visual system play its full role in visual function.

Taste and olfactory sensitivity

There has so far been very little investigation of taste and olfaction in the domestic chick, let alone in the embryo. Chicks were once thought to almost totally lack these two senses. However, they do have taste receptors (Kare *et al.*, 1957; Gentle, 1971) and their olfactory epithelium is reasonably well developed (Croucher and Tickle, 1989; and Fig. 1.8B). Ganchrow and Ganchrow (1985) have mapped the distribution of taste receptors in chicks at one day posthatching and found an average of over 300 taste buds in the oral cavity, mostly on the upper beak.

The nasal placode begins to form on day E2, and the nasal pits have elongated by day E4 (Barcroft and Bellairs, 1977; Wendel Yee and Abbott, 1978; Will and Meller, 1981). The olfactory axons start to grow on day E5 and have formed thick bundles by day E6 (Mendoza *et al.*, 1982). Within the brain, the olfactory structures form rapidly on days E4 and E5 (Tsai *et al.*, 1981). These structures include the olfactory bulbs, which are ready to receive the projections from the neurons of the olfactory epithelium, the olfactory tuberculum and the piriform cortex. On day E6 the upper beak has grown forward between the nostrils (Fig. 1.8A) and the presumptive olfactory epithelium has become localized in the roof of the nasal cavity and the adjacent septum, and the nasal capsule is beginning to form (Tamarin *et al.*, 1984; Croucher and Tickle, 1989). Although there is considerable development of olfactory structures early in the incubation period, the olfactory epithelium is not well-developed by day E10 (Okano and Kasuga, 1980).

In addition, during most of the incubation period, however, the nares are blocked by tissue and so it is unlikely that the embryo perceives olfactory stimulation until this tissue degenerates on day E20 (Romanoff, 1960), unless chemoreception occurs via reflux of fluids from the mouth onto the nasal epithelium. By day E20 the beak has penetrated the membranes to project into the air sac (Fig. 1.3) and breathing begins. Hence, it would seem that olfactory perception proper is delayed until the day before hatching. Tolhurst and Vince (1976) have tested sensitivity to odours in embryos at this stage of development. The top of the egg, around the air sac, was removed and responses such as beak-clapping, gaping, head-shakes, and heart rate were assessed when odorants were presented. Presentation of dichloroethane, formic acid and cineole produced an increase in heart rate and beak-clapping, and the first two of these odorants produced increased head-shaking.

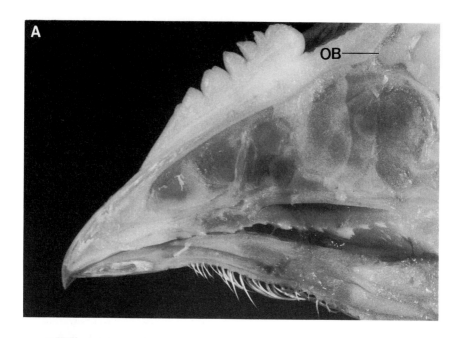

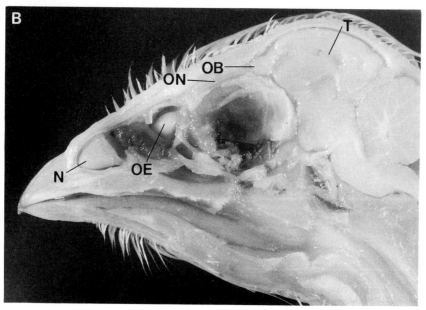

Presentation of amyl acetate did not lead to consistent results. Blocking the nostrils with wax abolished the behavioural responses to odours, indicating that they are indeed driven by olfactory sensation, and not by taste detected either via the taste buds in the mouth or the nerve endings of the trigeminal nerve at the back of the palate.

It is possible that the embryos would respond more strongly to natural odour mixtures, such as nesting material. In fact, hatched chicks have been shown to form an attachment to the odour of nesting material (Burne and Rogers, 1994a,b and Chapter 3). Imprinting to odours of the hen and nest may therefore occur prior to hatching, although this has yet to be tested. Alternatively, the embryo may respond to any, or all, of these chemicals via the trigeminal system for chemoreception, rather than via the olfactory system.

Little is known of taste sensitivity in the chick embryo, although imbibition of amniotic fluid starts to occur on day E11 or E12 (Romanoff, 1960). Later, just before membrane penetration when the amniotic membrane begins to break down, yolk sac material and allantoic fluid is ingested (Vince, 1977). It is possible that taste plays a role in this ingestive behaviour, but this is not known. Nor is it known whether any learned preferences develop from this behaviour, although learned taste preferences can occur after hatching (Gentle, 1972; and Chapter 3). Vince (1977) has shown that the taste system becomes functional at least one day before hatching, but it is likely to do so earlier unless the time of maturation of taste and olfaction are synchronized (Shapiro, 1981). Vince (1977) tested day E20 embryos, after the beak had penetrated the membrane into the air sac, by removing the air sac end of the egg and then using a paint brush to apply various solutions to the beak. Beak-clapping and heart rate were scored, and both were found to increase in response to the ingestion of sour, sweet or salt tasting solutions. Distilled water also produced a response, against the background control of egg-fluid. Blocking the nares with wax did not affect the results, indicating that taste rather than olfaction is involved.

Concluding Remarks

In this chapter I have outlined in chronological order the main events in the development of brain and behaviour of the chicken embryo. I have discussed

Fig. 1.8. (Opposite) Longitudinal section of the chicken head and beak showing the nasal cavities and olfactory epithelium. A: section with the septum (primarily a layer of cartilage) which separates the two nostrils still in place; this partition separates the two nostrils. B: section with the cartilage removed to reveal the olfactory epithelium (OE), olfactory nerve (ON), olfactory bulb (OB) in the telencephalon (T) and the structure of the nares (N). These preparations are from hatched chicks.

them without elaborating on how the sequence of development can be influenced by sensory stimulation, or how deprivation in one sensory system may be compensated for by developmental changes in another sensory system. These important issues are the topic of the next chapter. Each sensory system and each neural function may have its own pattern of timed development, but it does not differentiate separately from all the other systems and at all levels of neural organization.

It will be apparent that more is known of the development of the auditory and visual systems compared to the vestibular, tactile, proprioceptive, taste and olfactory systems. These latter systems are likely to be equally important to the avian embryo, and information processed by these sensory systems prior to hatching is likely to be essential to the newly hatched chicken.

On presently available information, it appears that the senses of audition and taste become functional at about the same stage of embryonic development (around day E12), whereas tactile sensation, proprioception and sensitivity to vestibular stimulation occur much earlier (day E6 to E10) and the visual and olfactory systems become functional much later (day E18 to E20). It is therefore interesting to speculate that there may be, as yet, unknown associations between the developing auditory and taste systems.

Study of the developmental events of the nervous system has highlighted the important role played by specific activities in neural circuits, and even of forming memories by the embryo. The findings have relevance to all species; even in humans essentially the same chronological order of development is followed (Gottlieb, 1973).

2

ENVIRONMENTAL INFLUENCES ON DEVELOPMENT OF THE EMBRYO

Summary

- Internal and external environmental influences on the development of the chick embryo are discussed.
- The influence of various forms of sensory stimulation has been studied either by depriving the embryo of particular sensory inputs or by providing it with specific forms of input at certain stages of development, followed by looking at the outcome on brain and behaviour.
- Tactile, vestibular and proprioceptive stimulation may synchronize hatching time in batches of eggs.
- Auditory stimulation may lead to the formation of specific memories, displayed by the chicks after hatching as preferences for familiar sounds. In fact, the chicks may imprint on the maternal call before hatching. The development of the auditory system is also influenced by self-stimulation by the embryo's own vocalizations, made during the last three or four days before hatching. Mutual communication between the embryos and the hen occurs from day E18 on. The embryo may also obtain influential stimulation from its own heart-beat and the heart-beat of the hen.
- Light stimulation early in development of the embryo accelerates development. Later in development it influences the development of the eye and thalamofugal visual projections. In particular, it causes the thalamofugal visual system to be organized asymmetrically.
- The level of sex hormones in the embryo alters the degree to which light stimulation affects the development of the thalamofugal visual projections.
- Olfactory stimulation and atmospheric pressure may also influence the development of the embryo, but less is known of these factors.
- Experiments investigating the intermodality effects of sensory stimulation indicate that the ordered sequence of development of the various sensory systems (as outlined in Chapter 1) is important for the normal development of behaviour.

Introduction

The development of the embryo is influenced by sensory stimulation. Indeed, once a given sensory system has become sufficiently developed to send inputs to the higher processing centres of brain and is connected to motor output pathways, it is possible to condition the embryo to stimulation via this sensory modality. We have already mentioned that the embryo can be conditioned to auditory stimuli as early as day E16, although this ability is improved on days E19 and E20 (Sedláček, 1964; Chapter 1). Conditioning in this context means that the embryo changes its responses to an auditory stimulus, indicating that it has formed a memory of that stimulus although this may not be a strong and persistent memory. Imprinting memory, by contrast, is stronger and more persistent, even to the extent of affecting behaviour after hatching. It is possible to imprint the embryos to auditory stimuli on day E19 or E20 (Grier *et al.*, 1967), and there is also some evidence that at these later stages of embryonic development it is possible to develop a preference for light of different wavelengths or simply for white light (Rajecki, 1974). The ability of the embryo to learn is not surprising as the chick is a precocial species and its survival depends, not only on being able to form essential memories immediately after hatching, but also learning before hatching. By imprinting on the maternal calls, the embryo plays its role in establishing the mother–chick bond necessary for maintaining proximity and thus survival.

It is not so easy to see what essential role the ability of the embryo to learn in the visual modality may have, particularly since only limited visual inputs can reach the embryo through the eggshell and membranes. Although light passes through the shell and membranes quite readily, patterned visual inputs, apart from temporal on-off patterning, cannot reach the embryo. It is, of course, possible that light stimulation of the embryo prior to hatching simply primes an already matured visual system so that it functions more effectively after hatching. That is, the exposure to light before hatching may simply stimulate the visual system non-specifically without causing learning prior to hatching, but facilitating visual learning after hatching. In support of this hypothesis, Dimond and Adam (1972) showed that chicks exposed to a flickering light prior to hatching were more likely to approach a flickering light, but not necessarily one flickering at the particular rate to which they had been exposed. The exposure to flickering light during embryonic development had apparently facilitated approach to flickering visual stimuli quite nonspecifically. There is, however, some evidence that the embryo may be able to learn about specific stimuli, particularly auditory stimuli (see pp. 47–51).

In experiments investigating visual imprinting in chicks it is customary to use chicks that have hatched from eggs incubated in darkness and to prime the hatched chicks by brief exposure to diffuse light before imprinting them

(Bateson and Wainwright, 1972; Horn, 1985). The priming enhances activity in the visual regions of the forebrain and facilitates approach during imprinting (Bateson and Seaburne-May, 1973). Furthermore, the imprinting is said to be stronger in those chicks that have been incubated in darkness and exposed to light early after hatching (Moltz and Stettner, 1961), possibly because chicks treated thus are less fearful of visual stimuli and so approach and follow them more readily. In fact, Dimond (1968) has reported that chicks hatched from eggs incubated in darkness are less likely to show the fear response of freezing in the presence of a moving visual stimulus than are those exposed to light during incubation. Under the laboratory conditions of dark incubation and light priming, the chicks have a well-defined optimal period of imprinting at around 15 h posthatching (Ramsay and Hess, 1954). In the natural situation the eggs are likely to have received light exposure throughout the incubation period, which may both prime the visual system for earlier imprinting after hatching and also influence the growth of the visual pathways.

In fact, stimulation of the sensory modalities can influence their development long before sensory inputs reach the neural centres for higher processing, when learning is possible. As already discussed, the sensory receptor neurons begin to differentiate as early as day E2 or E3 (Table 1.1), and thus it is potentially possible for sensory stimulation to direct or modify the differentiation of the neural connections. It has, for example, been shown recently that activity of the visual neurons early in their period of differentiation determines correct wiring in the visual system of the chick (Péquignot and Clarke, 1992a,b) and of the rat (Galli-Resta *et al.*, 1993). If the neural activity is blocked by the drug, tetrodotoxin, incorrect connections form. Growth of the visual projections to the forebrain of the chick is also dependent on light stimulation, as will be discussed in detail.

The question of whether sensory stimulation or, in fact, behaviour of the embryo itself could influence the maturation process has historically been one of great interest, and study of maturation of the chick embryo has been central in resolving the issue. As Gottlieb (1971a, 1976) so clearly delineated, the main issues in the development of behaviour were concerned with the question of whether certain behaviours are innate or acquired through experience. There were two main points of view. The predeterminist position claimed that during development genes determine the growth of bodily structures, including the nervous system, and, as maturation occurs according to a genetic programme, the genes have a primary role in causing the animal to behave in the way that it does. According to this hypothesis, behaviour patterns for survival have been selected by evolutionary processes, and therefore the species carries the programme for certain important behaviour patterns in the genome. Maturation was, therefore, considered to be a unidirectional process of expressing the genetic programme. The alternative position considered that, although genetic events give rise to the structural components of early development, continued maturation relies on functional interactions with the environment. That is, it was considered that there were bidirectional relationships between maturation, function and behaviour.

The latter hypothesis was linked by Gottlieb and Kuo (1965) to that of self-stimulation (see pp. 18–19). They drew attention to the fact that the embryo is subject not only to external stimulation but also stimulates itself through its own activities, motility and vocalizations. Throughout embryonic development, these might influence the maturation process.

According to the original view that behavioural development is pre-determined by the genetic programme, stimulating a developing animal, or depriving it of certain forms of stimulation, should have no effect on the development of behaviour, either in terms of its timing of occurrence, threshold for elicitation or the details of its patterning. The alternative view, however, predicted that sensory stimulation or deprivation during development would affect behaviour. Thus, the conflicting views could be experimentally resolved by studying the effects of stimulation or deprivation of developing chick and duck embryos (Gottlieb, 1971b). Given their development in the egg, it was simpler to conduct controlled experiments on these species than on mammalian embryos. Only now, some 20 years later, are the same issues being systematically addressed in mammalian species. The earlier studies with chick and duck embryos were pivotal in demonstrating the general principle that maturation of brain and behaviour is not exclusively predetermined, but relies on a continual interplay between genetic determinants and environmental inputs throughout the course of development.

Certain forms of stimulation during embryonic development induce change in behaviour, whereas others facilitate the process of maturation (Gottlieb, 1976). An inducing (or initiating) event leads to a change in behaviour, as in the case of the formation of an auditory preference by exposing the embryo to particular auditory stimulation. Facilitation events change the time course of development. For example, as will be discussed in more detail later, certain forms of stimulation can alter the timing of hatching. Light exposure can, for example, lead to earlier hatching (Lauber and Shutze, 1964), and exposure of quail embryos to the clicking sound made by other embryos retards the development of some eggs and accelerates the development of others so that they all hatch at the same time (Vince, 1973). Other demonstrations of the importance of sensory stimulation during embryonic life involve either exposing embryos to sounds during incubation or depriving them of auditory stimulation, even that produced by their own vocal apparatus, and investigating the outcome on responsiveness to sounds after hatching. For example, devocalization of ducklings, which prevents self-stimulation by vocalizations, delays development of their ability to discriminate their maternal call from that of another species of duck (Gottlieb, 1976; and pp. 48 and 49). Self-generated experience clearly influenced the processing of auditory information at some level of neural organization.

Sensory functioning must also be maintained by appropriate sensory stimulation. If, for example, a chick is raised in darkness for many days after hatching, it shows impaired visually guided pecking. The embryo has similar sensory requirements for maintaining sensory function, as will be illustrated in this chapter.

Bateson (1991) has introduced a fourth descriptive category for behavioural change during development. He has singled out events that permit changes in the patterns of behaviour but do not themselves produce those changes, and called them enabling events. An example of an enabling event occurring after hatching comes from an experiment by Cherfas (1977), who showed that exposure of chicks to patterned light (two black crosses) subsequently enabled them to learn to discriminate between the colour of two beads, one of which had been coated with a bitter tasting adversant (see p. 140). Enabling can also occur before hatching. Experiments by Lickliter, to be described later in this chapter, have illustrated that sensory experience of the embryo in one modality can enable appropriate learning in another modality.

Thus, development of the embryo depends on inducing, facilitating, maintaining and enabling events from the environment. All of these environmental inputs interact with the genetic programme for development. To recognize the complex, interactive role of environmental influences on brain and behavioural development does not imply that the developing embryo is essentially reactive to that stimulation. Conversely, nor is the behavioural development of the embryo determined largely by intrinsic (genetic) constraints. Development follows certain internal rules and self-correcting processes that are continually modified by environmental influences (for further discussion of this point see Bateson, 1976).

The discussion of sensory stimulation and development of the chick embryo must be largely confined to the auditory and visual modalities since there has not yet been consistent study of the other sensory systems in their own right. Nevertheless, some attention has been given to interactions between the effects of the various sensory systems on development, and this will be mentioned after each sensory system has been considered in isolation.

Tactile and Vestibular Stimulation

During the incubation period the hen periodically shifts the eggs in the nest. Such movement of the eggs must provide the embryo with stimulation of the vestibular system, as well as tactile and proprioceptive stimulation. Shapiro (1981) suggested that tactile stimulation of the embryo may prime tactile learning after hatching, and so facilitate posthatching social attachment. For example, tactile preferences may develop for the texture of feathers or the nest itself.

The variation in hatching time in domestic fowl eggs has been shown to be less in batches of eggs incubated in physical contact with each other than in those in which each egg is separated from the others (Oppenheim, 1973). The interval between pipping and hatching was altered somehow by communication between the eggs in physical contact with each other. This stimulation may be tactile, as movement of the embryos in the eggs may cause vibration, or it may be auditory, assisted by conduction via the fluids of the eggs and shell contact (see pp. 47–53 for effects of auditory stimulation on the development of the embryo). Vince and Chinn (1972) stimulated embryos from 48 h before the expected time of hatching by placing them in a cradle attached to a vibration generator (1.5 Hz) and found accelerated hatching, although there was some retardation of the posthatching age at which these chicks could stand up and walk compared to chicks hatching at the normal time. Despite the extended period from hatching to standing and walking, the chicks from the accelerated hatching condition still stood and walked earlier than chicks hatched from eggs which had not been stimulated if one estimated their age from the time of incubation or the 'expected' time of hatching. Apparently, the chicks that hatched earlier were slightly weaker and less well coordinated than those hatching later (Vince and Chinn, 1972). In fact, these researchers suggest that motor development after hatching may depend on the embryo having adequate tactile stimulation in the confined space of the egg before it hatches. Just before hatching the embryo almost completely fills the available space inside the egg shell. Its legs and feet are flexed and held in this position by the shell and must press against it when they move. This form of tactile and proprioceptive stimulation before hatching may be essential for the development of normal motor coordination after hatching.

Tactile, vestibular and proprioceptive stimulation might also interact with input in the other sensory modalities. In fact, Schneirla (1965) proposed that exposure of the embryo to quantitative aspects of tactile and proprioceptive stimulation (including rhythmical activity) may affect the later responses of the hatched chick to auditory and visual stimuli. In other words, there may be intermodality experience which shapes perception and may play a role in species-specific recognition (Gottlieb, 1973). There has, as yet, been little experimental evidence obtained in support of this interesting hypothesis about interaction between the developing somatosensory system and either the auditory or visual system, but recent work by Lickliter and colleagues (Lickliter and Stoumbos, 1991, and pp. 67–70) has demonstrated clearly that such intermodality interaction occurs between the auditory and visual systems.

We now know that there is no reason why an interaction between vestibular, tactile or proprioceptive inputs and either auditory or visual sensory information would need to be delayed until after hatching. Although the auditory and visual sensory systems mature after the vestibular, tactile and proprioceptive sensory systems, they do so before hatching (Chapter 1). Indeed, it is possible that the tactile, vestibular and proprioceptive stimulation that occurs with egg shifting plays an essential role in auditory learning by the embryo because the hen calls as she is shifting the eggs (Guyomarc'h, 1972; Impekoven and Gold, 1973). Similarly shifting of the eggs is accompanied by light stimulation (see p. 66) and thus vestibular, tactile and proprioceptive circuits may become associated with processing of visual information. All of these potential interactions deserve further investigation.

Auditory Stimulation

As the chick embryo begins to respond to sound from as early as day E12 of incubation, it is not surprising that its auditory system is well developed before hatching and that the embryo can even form memories of auditory stimuli. Rajecki (1974) exposed embryos to either white light or sound (on-off at one second intervals) from day E13 to E18 and found that the chicks displayed less distress calling, compared to controls incubated in silence and in darkness, when they were in the presence of a stimulus in the same modality as the one to which they had been exposed previously. Furthermore, the exposure to sound prior to hatching was more effective than similar exposure to light. This indicates the primacy of auditory perception and processing during embryonic development.

In fact, embryos appear to be able to form quite specific memories of auditory stimuli. Grier *et al.* (1967) exposed chick embryos to a patterned sound (200 Hz presented as 1 s on followed by 1 s off) from day E12 to E18 and found that, after hatching, the chicks approached the familiar tone in preference to a novel tone (2000 Hz presented in the same pattern). Controls, which had not received sound stimulation during incubation, had no preference for either tone. The chicks exposed to the sound pattern during incubation also followed a moving model which emitted that familiar sound for longer than did the controls. That is, they appeared to have imprinted on the sound presented prior to hatching (Chapter 3). The experimental design would, however, have been improved by including a group exposed to the 2000 Hz tone (Bolhuis and van Kampen, 1992). Nevertheless, a similar result was obtained by Simner and Kaplan (1977), who exposed embryos from day E12 to E19 to 500 or 1025 Hz tones and found entrained preferences at 8 h after hatching.

Species-specific auditory learning

Although newly hatched chicks preferentially approach a speaker emitting their own species maternal call even if they have not been exposed to it during incubation (Ramsay, 1951; Gottlieb and Vandenbergh, 1968), it is likely that their perception of their own mother's call can be sharpened by exposure to it prior to hatching. Support for this hypothesis comes largely from Gottlieb's studies using duck embryos (mentioned earlier). Although Gottlieb (1976) attempted to demonstrate the effect in chicks, he found the results too variable and so chose to present only the results for the ducks. The ducks were deprived of auditory stimulation prior to hatching and were even prevented from hearing their own calls by being devocalized by putting a glue-like substance, colloidin, over and around the syrinx (Gottlieb and Vandenbergh, 1968). The ducklings which had been auditorily deprived prior to hatching showed no preference for their species-specific maternal call when first tested after hatching, whereas the controls preferentially approached this sound. Later, however, the deprived ducklings did show a preference for the maternal call. Apparently, the self-stimulation usually provided by vocalizing prior to hatching either facilitates development of the preference or in some way shapes the template for later recognition of the maternal call. In a subsequent experiment, Gottlieb (1979) found that day E22 duck embryos which were aurally inexperienced responded (by bill clapping) to only the low frequency components of the maternal call, whereas aurally experienced ones responded to both the low and high frequency components. As the aurally inexperienced embryos develop the ability to respond to both the low and higher frequency components by the time they are two days posthatching in age, the auditory experience *facilitates* the developmental process. The ability to perceive, or respond to, a broader range of frequencies means that there is increased specificity of recognition of the maternal call. There is less chance of mistaking the maternal call of another species, such as the chicken, for the species-specific maternal call (Gottlieb, 1978).

Like ducks, chicks vocalize before hatching (Guyomarc'h, 1974). Some commence vocalizing as early as day E18 and all are able to vocalize by day E20 (Gottlieb and Vandenbergh, 1968). They make three distinct kinds of calls, characterized by Gottlieb and Vandenbergh (1968) as distress, contentment and brooding-like calls. The first two calls have subsequently been termed peeps and twitters. The brooding call is said to resemble the call made by hatchlings when they are in close contact and drowsy. In fact, the chick embryos emit this call most frequently during the hatching process.

Although Gottlieb (1976) said that his results with chick embryos were too variable to warrant reporting, Kerr *et al.* (1979) demonstrated that auditory deprivation of chick embryos delays the sharpening of auditory perception that usually occurs during development without auditory deprivation. They deafened both ears by filling them with a wax compound on day E18. The hatched chicks were allowed 25 min of normal sound exposure after removal of the ear plugs before they were tested for auditory generalization on day three or four. Each chick was first habituated to a 800 Hz tone and then tested with five frequencies ranging from 800 to 1,000 Hz. Calling in response to a perceived novel sound was scored. The chicks which had been

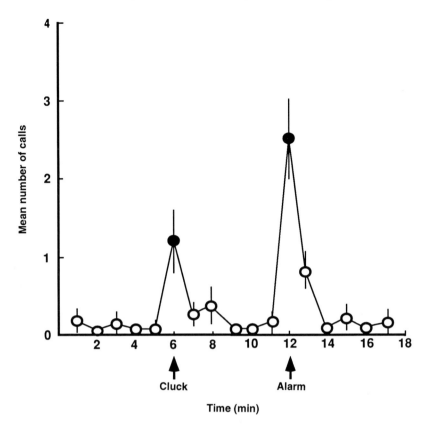

Fig. 2.1. Increased vocalization by day E19/20 chick embryos in response to hearing the hen's cluck or alarm call. The number of vocalizations (calls) made by the embryo was scored for each minute over a 17 min period. The hen's calls were played to the embryo intermittently during minutes 6 and 12. A total of 16 embryos were tested and the data presented are mean values with standard errors. Standard errors are indicated only where they exceed the size of the symbol. Note the significant elevation of vocalization by the embryos during the presentation of the hen's calls, and for 1 or 2 min after their presentation.

auditorily deprived showed flatter generalization curves than controls of the same age (three to four days). Their curves were, in fact, like those of control chicks tested on day 1 posthatching, suggesting that the auditory deprivation had delayed maturation of the auditory system.

Chick embryos exposed to the maternal call of their species do in fact respond by increasing beak-clapping and by vocalizing. Gottlieb (1965a) tested the responses of day E20 embryos to the maternal call over a 30 s interval. The call was played at intervals of 1.5 s. Beak-clapping increased markedly during each burst of sound, whereas increased numbers of vocalizations were made by the embryos in the silent period immediately following each presentation of the maternal call.

This result is most reliable. It has been repeated many times in class experiments; the data for one of these is presented in Fig. 2.1. Two maternal chicken calls were presented, a cluck and an alarm call, at regular intervals over a one-minute period. The number of vocalizations made by the embryo per minute was scored over the five minutes prior to presenting the cluck call, the five minutes between presentation of the calls and for the final five minutes after presentation of the alarm call. Note the large increase in vocalization rate during the one minute periods over which the calls were presented, and that the increase was greater in response to the alarm call. The latter result may, however, merely reflect intensity differences between the calls, since these were not balanced.

Vince (1980) exposed day E19 bantam chick embryos to the hen's alarm call and found a change in heart rate to be a more reliable response than vocalizing. Heart rate increased in response to the alarm call, but with repeated presentations of the call this response dwindled. Apparently, the embryos had habituated to the call. Furthermore, this diminished responding extended into posthatching life. At two hours after hatching chicks exposed to the call prior to hatching showed decreased eye opening and decreased vocalizations in response to hearing the call, compared to chicks not previously exposed to the call. Surprisingly, Vince (1980) found that as little as 55 s exposure to the alarm call on day E19 of incubation had a significant effect on responding after hatching.

Fält (1981) has demonstrated that chicks brooded and raised by a hen will preferentially approach the clucking call of the familiar hen rather than the clucking of an unfamiliar hen. On day 1 posthatching one third of the chicks choose the familiar clucking call, two thirds remained immobile ('no choice') and not one approached the unfamiliar clucking call. By day 4, 68% approached the familiar call, 32% made no choice and, again, not one approached the unfamiliar call. The increasing approach to the familiar call could reflect either learning in the posthatching period or increased locomotor ability, but most likely both of these. Chicks hatched from eggs incubated artificially until pipping and then placed with a hen, all made no choice between the calls on day 1 posthatching, but on day 4 there was a strong preference (89%) for the familiar hen's call. Chicks incubated artificially and placed with the hen at hatching showed a slight preference for the familiar call by day 4 posthatching, but this preference (29%) was much weaker than in the other two groups. Apparently, experience of the hen's clucking call prior to hatching greatly enhances the preference for that call. This does not imply that there is no further learning about the hen's vocalizations after hatching. In fact, Graves (1973) stated that chicks have a sensitive period for learning maternal calls that extends to day 4 posthatching, and Russock and Hale (1979) found that incubator-hatched chicks respond to the food call of a hen up until three to five days posthatching. Thus, within approximately the first four days after hatching the chicks must hear the hen's vocalizations if they are to retain responsiveness to them.

Vocal communication between the embryos and the hen

There is two-way communication between the embryos and the hen. The vocalizations produced by chick embryos also elicit responses from the hen. Tuculescu and Griswold (1983) monitored vocal interactions between chick embryos and the hen from day E18 of incubation to hatching. They categorized the embryos' calls into distress and pleasure calls. The distress calls included 'phioo', soft peep, peep and screech calls, whereas the pleasure calls included twitters, food calls and huddling calls. The vocalizations of the hen recorded were cluck, intermediate, food and mild alarm calls. They found a definite increase in vocalizing by both the embryos and the hen as hatching approached. The first calls of the embryos recorded were distress calls. As hatching approached screeches and soft peeps became more frequent, whereas peeps remained at a low level. The embryos did not emit pleasure calls until some three to five hours before hatching, and coincident with this the hen began to vocalize; chiefly with intermediate and cluck calls. The embryo–hen pattern of calling at two hours before hatching was analysed in more detail, and it was revealed that the embryos made pleasure calls mainly within 15 s of the start of an action by the hen. Distress calling by the embryos was not similarly linked to behaviour of the hen. In fact, twitters were emitted by the embryos following turning of the eggs by the hen and as she resettled on the nest, but the same acts suppressed food calls. Vocalizations by the hen, particularly the cluck and intermediate calls, were followed by the embryos making food calls, along with the inhibition of screeches and soft peeps. Thus, it can be concluded that the embryos respond to acts of the hen.

The converse also occurs: the hen responds to vocalizations of the embryos, particularly as hatching approaches. The embryos' calls stimulate the hen to turn the eggs, to cluck, to peck and to beak clap. Tuculescu and Griswold (1983) found that the calls of the embryos most likely to cause responding by the hen were screeches, whereas twitters and food calls were more likely to inhibit the responses of the hen.

To summarize, the embryos' distress calls elicit responses from the hen, whereas their pleasure calls follow those responses by the hen. The pattern appears to be identical to that found after hatching. It appears that the embryo may distress call when, for example, it gets cold. The hen then moves the egg in the nest and the embryo immediately pleasure calls. In fact, the most effective signals of the embryos are the intense distress calls, the peeps and screeches. These patterns of interaction become better synchronized as hatching approaches. There is clearly mutual communication between the embryo and the hen just prior to hatching.

Communication between embryos

Moreover, the embryos also communicate with each other. Chick embryos produce clicking sounds in the period between pipping the egg shell and hatching (Vince, 1966). In fact, the clicks stop abruptly just before hatching, but begin again shortly after hatching and continue for several hours (McCoshen and Thompson, 1968). The sounds depend on inspiration and expiration, but they are not produced by the passage of air through the glottis (McCoshen and Thompson, 1968). Instead, they may be caused by movement of the cartilage in the region of the glottis. Vince (1964, 1966) has demonstrated that in quail these clicking sounds synchronize hatching between members of the same clutch. In fact, the development of an isolated egg can be accelerated by exposing it to clicking sounds from a loudspeaker (Vince, 1973). The stimulated embryos hatch earlier.

Clicking serves the same purpose in embryos of the domestic fowl; exposure to clicks at the rate of three per second accelerates hatching time, whereas stimulation at the rate of 100 clicks per second has no such effect (Vince *et al.*, 1970). In the embryos which have been accelerated the weight of the hatching muscle increases earlier than in unstimulated controls (Vince, 1973).

So far there has not been any investigation of whether the clicking sounds made by a single embryo provide self-stimulation which, like vocalization, may facilitate the development of the auditory system. It might also interact with the effects of stimulation in another sensory modality.

Embryos also produce bursts of sharp, short signals of relatively high frequency by clapping the beak. Beak clapping begins between day E12 and E16 (Kuo, 1932a,b) but may not become audible until just before breathing begins (Freeman and Vince, 1974). It occurs at the rate of three or four per second. These signals are also likely to play an important role in synchronizing development and thus hatching time.

The embryos may also be exposed to low-frequency sound as they are knocked together when they are turned by the hen (Gottlieb, 1968; Impekoven and Gold, 1973). Low frequency (non-clicking) sounds are also produced by the embryos themselves, before they start to produce clicking sounds, and these retard the development of more advanced embryos, whereas the clicking sound accelerates the rate of less advanced embryos (Vince, 1970). In Japanese quail, Vince and Cheng (1970) have shown that the interval between the commencement of lung respiration and hatching is either lengthened or shortened by contact of the egg with other less advanced or more advanced eggs.

Heart-beat stimulation

The embryo may also obtain stimulation from its own heart-beat and/or that of the hen. Salk (1962) suggested that human infants may imprint on the maternal heart-beat, which provides a repetitive input throughout uterine development. In support of this hypothesis, Salk demonstrated that exposure of human infants to a heart-beat sound has a calming effect. The issue was taken up by Simner (1966), who hypothesized that the developing chick embryo might be similarly affected by exposure to the sound of its own heart-beat. The evidence which he has presented in support of this hypothesis is, however, tenuous. Following the idea of Kuo (1932a,b), he attempted to draw a relationship between the rate of heart-beat prior to hatching and the neuromuscular mechanisms concerned with vocalization (Simner, 1966). In other words, the two rhythms might have become entrained during embryonic development. Although he was able to show that vocalization by the chick is followed by a decrease in heart rate, there is no evidence of a causal relationship between the two phenomena.

In a subsequent study, Simner and Kaplan (1977) tested the preferences of hatchling chicks for heart-beat sounds. In one experiment the chicks were given a choice of recorded fetal heart-beat and a square wave white noise signal matching the pulse rate. Contrary to the experimenters' prediction, the chicks approached the latter in preference to the heart-beat recording. It is possible, however, that the square wave represented a supernormal stimulus, preferred over the actual heart-beat. If so, this result would lend support to the cardiac self-stimulation hypothesis. Further investigation of this matter would be of interest.

Lateralized auditory stimulation

One might extend the hypothesis to argue that the known dominance of the left hemisphere of the chick in auditory habituation learning is entrained by lateralized stimulation of the embryo's ears by auditory stimuli. Auditory habituation learning in chicks is disrupted by drug treatment of the left, but not the right, hemisphere (Rogers and Anson, 1979; Howard *et al.*, 1980). As yet, there has been no investigation of the possible role of lateralized auditory input in establishing this asymmetry in the forebrain, but it should be recognized that during the last stages of incubation the embryo's head is twisted into a position which places the left ear next to the thorax, whereas the right ear faces away from the body (Freeman and Vince, 1974). Although it is recognized that sound may be conducted from one ear to the other by both bone and air sac conduction, it is possible that the left and right ears receive relatively different stimulation. The left ear may receive a greater degree of self-stimulation by the heart-beat whereas the right ear would receive relatively greater stimulation from sounds produced externally to the egg. The latter would include the heart-beat and vocalizations of the hen in addition to unpredictable stimulation by sounds in the environment. If so, the lateralization of auditory processing in the chick forebrain would be generated in the same way as is lateralization for visual processing (see pp. 54–57).

Visual Stimulation

As already mentioned, brief exposure of embryos to bright light from a photo-lamp leads to changes in motility as early as day E3 (Chapter 1). Light exposure during incubation also has general effects on the rate of development. Several researchers have reported that illumination of the eggs reduced the incubation period, with no effects on hatchability (Shutze *et al.*, 1962; Lauber and Shutze, 1964; Gold, 1969; Siegel *et al.*, 1969). Lauber and Shutze (1964) found that the most pronounced effect of the light stimulation is during the first week of incubation, and that incandescent light is more effective than fluorescent light. In fact, they found that exposure of the eggs to fluorescent light during the second week of incubation retarded hatching. Since the experimental procedures controlled for temperature effects, the results were apparently due to the light exposure *per se*. Fluorescent light differs from incandescent light in that it has more of the blue range of the spectrum, it flickers and it has different intensity levels across the wavelength spectrum. Any one, or even all, of these factors may be responsible for the different effects of these two forms of light exposure.

Asymmetry of light exposure

Asymmetrical development of the eyes

This light exposure also affects the anatomical development of the eye. In chicks that have received the light stimulation, the right eyeball becomes larger and heavier than the left (Lauber and Shutze, 1964). It appears to be the light stimulation *per se* causing the enlargement of the right eye because the right eye of the embryo receives more exposure to light both early and late in development (Figs 2.2 and 2.3). On day E3 the embryo turns its head to the left, resulting in the left eye being placed against the yolk sac and the right eye being relatively more exposed next to the membranes and shell. Until day E14 the embryo remains with its left side and eye against the yolk sac and the right eye exposed. Some light can, of course, penetrate the shell membranes and yolk sac to reach the left eye, but it would be quantitatively less than the amount reaching the right eye, and also qualitatively different in terms of wavelength.

From day E14 to E17, re-orientation of the embryo in the egg leads to both eyes being enveloped by the yolk sac, until on day E17 or E18 the yolk sac retracts, again revealing the right eye. At this stage the left eye is occluded by the body, which would provide even greater occlusion of the left eye from light stimulation. Although Hamburger and Oppenheim (1967) have stated that, once the embryo has adopted the hatching position, the head is tucked under the right wing 'which covers the eye completely' (p. 174), at this stage rapid, whole body movements, termed 'startles' or type II movements, occur

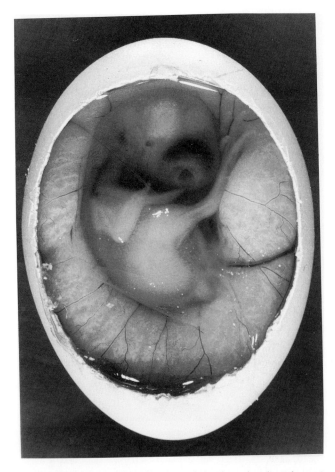

Fig. 2.2. A day E8 embryo. Note that it is lying on its left side so that the right eye is exposed.

at regular intervals and the wings rotate at the shoulder joint beginning with an upward and forward motion. These movements of the wing occur without parallel movement of the head, and so expose the right eye at least intermittently. Alternatively, the head moves without the rest of the body including the wing, thus also exposing the right eye. My own observation, as well as that of Bakhuis and van de Nes (1979), is that the right eye is exposed for considerable periods of time. In fact, the wing is not always in a resting position that exactly covers the eye (Fig. 2.3). Even if the eyelid is closed, light will reach the eyeball itself because the lid is transparent. Nevertheless, eye opening and closing occurs frequently from day E17 on (Freeman and Vince, 1974). In fact, light stimulation itself between days E18 and E20 causes an increase in the rate of opening and closing the eyelid (Kuo, 1932a).

Fig. 2.3. A day E20 embryo with the shell and membranes of the air sac removed. Note that the embryo's head is turned to its left side so that the left eye is occluded by the body, and the right eye is exposed, apart from partial covering by wing feathers depending on the position of the wing. There was no disturbance of the wing feathers caused by removing the membranes in this preparation.

Thus, asymmetrical stimulation of the eyes by light may cause asymmetry of eye size. As yet, the ages at which light stimulation may affect eye growth have not been delineated. Incidentally, it is possible that the earlier pigmentation of the right eye, mentioned in Chapter 1, depends on the asymmetrical light-stimulation, although this is a very early event in development.

The asymmetry of eye structure becomes manifest in the newly hatched chicks as relative differences in focusing ability. Even at four weeks of age the right eye is relatively less hypermetropic (far sighted) than the left (Noller, 1984). Not only is the resultant asymmetry in refractive ability of the eyes likely to have consequences for visual perception in the left and right monocular fields (see pp. 87–90 and 104–108), but it may influence development of the visual connections of either eye. In addition, because the right eye of the embryo receives more light than the left, the axons in the optic nerve connected to the right eye may myelinate sooner than those connected to the left. Although this has not been demonstrated in the chick embryo, there is clear evidence that in rabbits the right eye opens first and that this leads to myelination of the right optic nerve fibres in advance of the left (Narang, 1977; Narang and Wisniewski, 1977). Moreover, by artificially opening either eye, the light stimulation itself has been shown to be the factor which accelerates the myelination of fibres from the right eye (Tauber *et al.*, 1980). Thus, for a finite time during development the higher neural processing regions are likely to receive differential inputs from either eye, which, in turn, may influence the development and functioning of these centres in a lateralized manner.

Asymmetrical development of the visual projections

There are definite effects of light stimulation on the development of the second order visual projections from the thalamus (OPT region; see Chapter 1, Fig. 1.5) to the hyperstriatal region of the forebrain of the chick. In chicks hatched from eggs which have been exposed to light for the last three days of incubation, there are more projections from the left OPT, which receives input from the right eye, to the forebrain than from the right (Rogers and Sink, 1988; Fig. 2.4). The asymmetry is located primarily in the projections which cross the midline to project to the contralateral side of the forebrain (Boxer and Stanford, 1985; Fig. 2.4). There are more projections from the left OPT to the right hyperstriatum than vice versa. Light intensity of around 100 lux has been found to be sufficient to cause this effect, but so far there has been no attempt to test lower intensities.

The role of light stimulation during the last few days of incubation in the development of these thalamofugal projections was demonstrated by withdrawing the embryo's head from the egg on day E19 or E20, after tucking and membrane penetration had occurred, followed by occluding either the left or right eye with a patch and exposing the other to light (Rogers and Sink, 1988). The patch was removed at hatching, and from then on both eyes received light stimulation. The organization of the thalamofugal projections to the hyperstriatum was determined by injecting fluorescent tracer dyes into the HA on day 2, followed by sectioning the brain and counting the number of labelled cell bodies in the thalamus on day 4. The dyes injected into HA diffuse into surrounding regions including HD and IHA, the regions in which the thalamofugal visual projections terminate (Fig. 2.4). The neurons projecting to the site of the dye injection take up the dye and transport it retrogradely to label their cell bodies. In this case the cell bodies labelled are in the thalamus (OPT region) both ipsilateral and contralateral to the site of injection. For each dye, the ratio of the number of labelled ipsilateral to

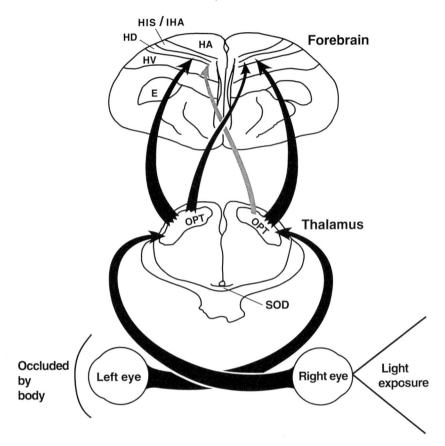

Fig. 2.4. The thalamofugal visual system of the chick. Axons of the ganglion cells in the retina of each eye project to the contralateral thalamus, in the region known as the nuclei opticus principalis (OPT). From OPT there are projections to both the ipsilateral and contralateral sides of the forebrain. These terminate primarily in the hyperstriatum dorsale (HD), and there are projections from HD to the hyperstriatum accessorium (HA). The contralateral projections from OPT actually cross the midline in the supra-optic decussation (SOD), rather than as drawn in the figure for simplification. Note that in young chicks there are more projections from left OPT to right HD than from right OPT to left HD. Other abbreviations: HIS, hyperstriatum intercalatum supremum; IHA, nucleus intercalatus hyperstriati accessorii; HV, hyperstriatum ventrale; E, ectostriatum.

contralateral cells is calculated to allow for variations in the amount of dye injected (Fig. 2.5). By injecting a dye which fluoresces blue under ultraviolet light into the HA on one side of the forebrain and another which fluoresces either white or yellow into the other side, it is possible to determine the asymmetry within a single brain (Rogers and Bolden, 1991). We use the dyes Trueblue and Fluorogold.

In the chicks which had received exposure of the left eye to light and right eye occlusion the asymmetry was reversed: there were more projections from

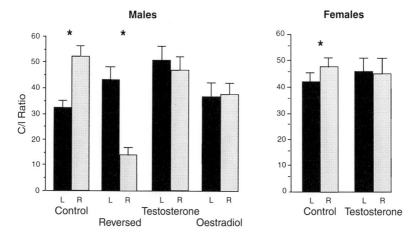

Fig. 2.5. This figure illustrates the presence of asymmetry in the thalamofugal visual projection of control male and female chicks and its absence following treatment of the developing embryos on day E16 with either testosterone oenanthate (12.5 mg) or oestradiol oenanthate (1.5 mg) injected into the egg. The number of ipsilateral (I) and contralateral (C) projections from each side of the thalamus to the hyperstriata in each hemisphere of the forebrain has been determined by injecting fluorescent dyes into the latter regions (see text). The mean C/I ratio (with standard errors) expressed as a percentage is plotted for injections into the left hyperstriatum (L, black bars) and right hyperstriatum (R, grey bars). Note that in controls the C/I ratio is significantly lower following injection of the left hyperstriatum than it is following injection into the right hyperstriatum. The asterisks indicate significant left–right differences calculated using paired comparisons. Thus, the asymmetry in control females is significant although this is not obvious from the means plotted as they do not illustrate individual differences. The data are collectively from Rajendra and Rogers (1993), Rogers and Rajendra (1993) and Schwarz and Rogers (1992).

the right OPT, which receives its input from the left eye, to its contralateral hyperstriatum. The chicks which had the left eye occluded, and so mimicked the normal situation, retained the normal pattern of more projections from the left OPT.

As could be predicted, chicks hatched from eggs incubated in darkness have symmetry of the thalamofugal visual projections (Rogers and Bolden, 1991). This symmetry occurs irrespective of whether the chicks are reared after hatching in darkness or in the light. Therefore, provided that both eyes of the embryo simultaneously receive the same degree of light stimulation the thalamofugal visual projections develop to an equal extent (for a summary of the role of light in the development of this asymmetry see Rogers and Adret, 1993).

It should be noted that, although it is during the last stages of embryonic development when the light causes the development of asymmetry, the asymmetry cannot be detected until after hatching. Injection of the dyes into day E20 embryos fails to reveal an immediate effect of the light on asymmetry (Rogers *et al.*, 1994a). Therefore, the light exposure triggers changes in the development of the neurons which continues into the posthatching period. In fact, the asymmetry persists throughout the first two to three weeks of posthatching life, after which time the development of the projections fed by the left eye has apparently caught up (Rogers and Sink, 1988; Rogers, 1991; see Chapter 4).

There are subcellular asymmetries in the chicken brain that may also be established by the asymmetrical stimulation of the eyes by light prior to hatching. For example, in two-day-old chicks the density of synapses in the right HA is significantly higher (by 22%) than in the left (Stewart *et al.*, 1992b). Since IHA projects to HA (Dubbeldam, 1991), the asymmetry in synaptic density may reflect the greater number of contralateral projections from left OPT to right hyperstriatum. The consequences of such asymmetries may be widespread and influential in visual functioning, particularly since HA is the main source of the septomesencephalic tract, which projects back to retinorecipient cells in the thalamus (Reiner and Karten, 1983; Micelli *et al.*, 1987). Via this feedback loop, the synaptic asymmetry in HA might further modulate visual inputs asymmetrically.

Interaction between light and hormonal condition

The hormonal condition of the embryo during the last few days of incubation also influences the development of the thalamofugal visual projections (Schwarz and Rogers, 1992; Rogers and Rajendra, 1993). If either of the sex hormones, testosterone or oestrogen, is injected into the egg on day E16 in a slow release form, which leads to elevation of the hormone level until after hatching, no asymmetry develops in response to asymmetrical light stimulation of the eyes (Fig. 2.5). In other words, the asymmetry of the thalamofugal visual projections occurs in response to light stimulation during the last stages of incubation only if the level of the sex steroid hormones is low. Higher levels of either of these hormones appears to stimulate growth of the visual projections from both sides of the thalamus and so masks the effect of asymmetrical light stimulation (Rogers and Rajendra, 1993). Thus, the levels of circulating sex hormones presumably cause the known difference in the degree of asymmetry in males and females: female chicks have a lesser degree of asymmetry than do males (Adret and Rogers, 1989; Rajendra and Rogers, 1993).

Throughout the incubation period the oestrogen levels of female embryos are considerably higher than those of males. Synthesis of 17β-oestradiol begins on day E4, and in females the plasma level continues to rise gradually until day E13, at which stage it increases markedly until day E17 (Woods and Brazzill, 1981). From day E17 to day E20 the plasma level of 17β-oestradiol falls slightly, and then begins to rise again after hatching (Tanabe *et al.*, 1979). In male embryos the 17β-oestradiol concentration in the plasma is two- to threefold lower than in female embryos, and it remains relatively

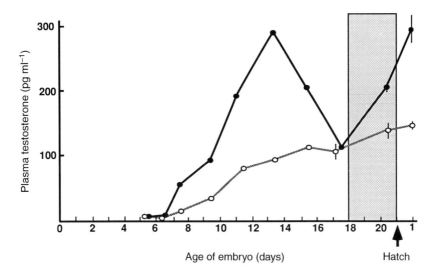

Fig. 2.6. Mean plasma testosterone levels (pg ml⁻¹) are plotted for male (black circles) and female (white circles) chicks at various ages from day E5 of embryonic development to day 1 posthatching. Standard errors are indicated only where they exceed the size of the symbol. The grey area indicates the stages of development during which light stimulation interacts with hormonal condition to influence the development of the thalamofugal visual projections (see text). The data have been constructed from Tanabe *et al.* (1979) and Woods *et al.* (1975).

constant throughout the embryonic period (Woods and Brazzill, 1981; Gonzales *et al.*, 1987). The higher levels of oestrogen in female embryos might explain why the development of their thalamofugal visual projections is less responsive to the asymmetrical light stimulation during the last days before hatching, but we must also consider the levels of testosterone.

Synthesis of testosterone by the embryo begins on day E5 and the levels circulating in the plasma are approximately the same for both sexes until day E7, when the levels in males exceed those in females (Woods *et al.*, 1975). In both sexes, however, there is an increase in the plasma testosterone concentration until day E13 in males and day E15 in females. In males, there is a marked drop in the plasma testosterone level from day E13 to E17, and then the level rises again on day E20 (Fig. 2.6). The pattern of testosterone secretion in females in this period just before hatching is less clear. Woods *et al.* (1975) report a decline in the level from day E15 to E17, but the value for day E17 is not consistent with that reported by Tanabe *et al.* (1979) for females of the same age. In Fig. 2.6, an average of those two disparate levels has been plotted for day E17, producing an impression that plasma testosterone levels rise gradually in females throughout the entire incubation period. Further information is needed to clarify the situation, but the marked trough in plasma testosterone levels in males on day E17 would seem to be genuine.

Thus, over the sensitive period for the effects of light on the development of the thalamofugal projections, there is little or no difference in the levels of testosterone between males and females, coupled with severalfold higher levels of oestrogen in females compared to males. The net effect is that the female embryos have higher circulating levels of the sex steroid hormones, and this provides a convincing explanation as to why asymmetrical light stimulation prior to hatching causes a lesser degree of asymmetry in the thalamofugal visual projections of females. It should be noted, however, that, since some degree of asymmetry of the thalamofugal visual projections is present in females, the circulatory levels of oestrogen plus testosterone in females must be at a level which still allows light to have at least some effect on the development of these visual projections. Progesterone levels are also higher in females (Tanabe *et al.*, 1979) and this may possibly counteract the action of testosterone to some extent, or directly affect development of the visual projections by acting in its own right. At present, it is difficult to say anything directly about the role of progesterone in development as the serum levels are highly variable and show no overall trends for change during development (Gonzales *et al.*, 1987).

One report has provided evidence that the source of testosterone in avian embryos may not be entirely the gonads but also the maternal bird (Schwabl, 1993). In canaries, the testosterone content of the egg yolk was found to increase with order of laying. A similar situation may occur in chicken eggs. It is known that in chickens the testosterone level in the plasma of the hen begins to rise at around 8 h before ovulation of the follicle and peaks at 4–5 h before ovulation. Since the egg is laid 24 h after ovulation and the next follicle ovulates within 15–75 min after laying, each successive egg may be exposed to a higher level of testosterone, which passes from the maternal plasma into the yolk sac. Thus, laying order may influence the level of testosterone to which the developing embryo's brain is exposed with consequences on brain development and behaviour after hatching (Winkler, 1993). In the context of the present discussion, it is possible that laying order may influence the degree of asymmetry in the thalamofugal visual projections, depending on how long any influence from the maternal source of hormone persists during incubation. This deserves attention in future research.

As has been discussed, light stimulation from the environment and internal hormone levels interact to influence the development of the thalamofugal visual projections. The outcome affects behaviour after hatching, as will be shown in Chapters 3 and 5. As yet, there has been no attempt to look for asymmetry in the tectofugal visual system of the chick, but asymmetry has been found in this visual pathway in adult pigeons (Güntürkün *et al.*, 1989) and this asymmetry is also determined by asymmetrical stimulation of the eyes by light prior to hatching (Güntürkün, 1990). The asymmetry is present in the neurons in layers 2–13 of the tectum. The average surface area of cells on the right side, compared to the left, is larger in layer 13 and smaller in layers 2–12 (Melsbach *et al.*, 1991). This is the case for light-reared, but not dark-reared, pigeons (Güntürkün, 1993; see Bradshaw and Rogers, 1992, pp. 76–85 for a summary). It therefore appears that both visual systems may have asymmetry dependent on light stimulation prior to hatching. Whether sex hormone levels also affect the development of the tectofugal visual system is not yet known.

Functional lateralization established by light exposure

There are also functional asymmetries in the chick brain that are similarly determined by exposure of the embryo to light prior to hatching (Rogers, 1982b). Remarkably, as little as 2 h of light exposure on day E19 is sufficient to establish brain lateralization in which the left hemisphere of the forebrain (and right eye) plays a dominant role in learning to categorize grains of food as different from small pebbles adhered to the floor, and for the right hemisphere to play a role in activating copulation and attack behaviour (Rogers, 1982b; Zappia and Rogers, 1983). These hemispheric differences in function will be discussed in detail in subsequent chapters (Chapters 3, 4 and 5). Here it is simply important to point out that chicks hatched from eggs that have received light prior to hatching learn to categorize food grains as different from small pebbles adhered to the floor (see Fig. 3.7) better when they are tested monocularly using the right eye than when they use the left eye. As for the asymmetry in the thalamofugal visual projection, it is possible to reverse completely the direction of the functional asymmetry so that chicks using the left eye perform better on the pebble–grain categorization task than do those using the right eye. This reversal is achieved by withdrawing the embryo's head from the egg on day E19 or E20, applying a patch to the right eye and allowing the left eye to receive stimulation by light (Rogers, 1990). Controls that have experienced the same operative procedures but with the left eye patched rather than the right, resulting in a situation which mimics the normal state, retain their superior performance when using the right eye.

Hormonal condition prior to hatching also affects the asymmetry for pebble–grain categorization. Females have less asymmetry of monocular performance of the pebble floor task than do males (Zappia and Rogers, 1987), a result that correlates with their lesser degree of structural asymmetry in the thalamofugal visual projections (Rajendra and Rogers, 1993). At least some of the other sex differences in behaviour between males and females in early posthatching life may similarly result from variations in the asymmetry of these visual projections, but confirmation of this prediction awaits further research.

Exposure of the embryo to light for a very brief period on day E19 establishes (at group level) both structural and functional asymmetries in the brain. Chicks hatched from eggs incubated in darkness have no asymmetry in a consistent direction within the group, although some individuals may be lateralized in one direction and others in the other direction so that no group bias is evident.

Chicks hatched from eggs incubated in darkness from day E17 to hatching form a more flexible group structure than those hatched from eggs which have been exposed to light during this period (Rogers and Workman, 1989). The chicks hatched from eggs exposed to light formed a more rigid social hierarchy for gaining access to food, scored over the first two weeks of life, whereas those hatched from eggs incubated in darkness over the sensitive period from day E17 to hatching showed more day-to-day variability in the rank order (for details see Rogers and Workman, 1989). Consequently the lowest ranking chicks of the dark-incubated groups were more successful in competing for food than their counterparts hatched from eggs exposed to light (Fig. 2.7). Furthermore, the dark-incubated chicks tended to gain access to the food dish by climbing over the top of the other chicks, whereas those hatched from eggs exposed to light pushed their way into the food dish from underneath, between the legs of the other chicks already feeding at the dish.

Although Rogers and Workman (1989) reasoned that these differences in competitive group behaviour might be linked to the presence or absence of asymmetry in a consistent direction within the group that has received exposure to light, there is so far no direct evidence for such a link. As they proposed, it is possible that the hierarchy might be more stable in a group composed of individuals all having asymmetry in the same direction, because social interactions might rely on more predictable responses (e.g. an increased likelihood of being attacked if a dominant animal is approached from its left side). On the other hand, Dimond (1968) has claimed that chicks hatched from eggs incubated in darkness are less fearful on the basis that they show less freezing behaviour when presented with a moving visual stimulus, compared to chicks exposed to light during the last week of incubation. The chicks hatched from eggs incubated in darkness also followed a moving imprinting stimulus more than those exposed to light during incubation.

Reduced fear might explain why the lower ranking individuals in groups composed of dark-incubated chicks are more likely to attempt to compete for access to the food, thus generating a more variable social structure. Yet, there is some evidence that control of fear responses is lateralized to the right hemisphere (Phillips and Youngren, 1986) and therefore it too may be dependent on the presence or absence of light-determined lateralization. Further investigation of these potential links between neural organization and behaviour is warranted.

Irrespective of the manner in which light stimulation before hatching operates, it has broad effects on behaviour after hatching. Therefore, one might question whether eggs incubated under natural conditions by the hen receive sufficient light exposure to show these effects on structural and functional development, and this certainly deserves further study. However, as already discussed, during this critical stage just before hatching the

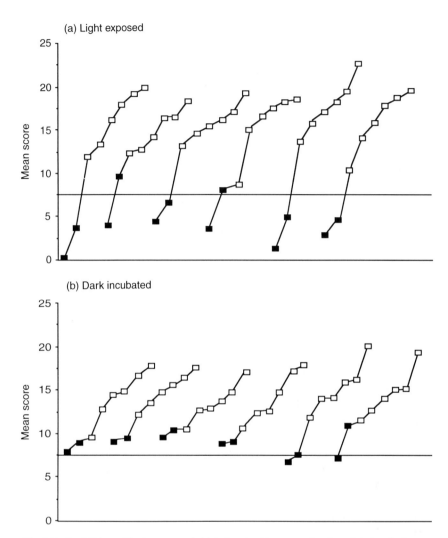

Fig. 2.7. Social hierarchies in groups of chicks hatched from eggs incubated during the last days of embryonic development either in the light (a) or in darkness (b). The rank orders have been determined in groups of eight chicks competing for access to a dish of food after being deprived of food for three hours. The number of times each chick gained access to the food dish during a 10 min testing period was scored daily from day 8 to day 16 of life. Thus a mean score ± standard error could be determined for each individual and this is plotted on the vertical axis. Each curve represents a group of chicks (6 light-exposed and 6 dark-incubated groups) and each point an individual's mean score (days 8 to 16), ordered according to rank from lowest ranking to highest ranking. The two lowest ranking individuals in each group are represented by solid black squares. The line at the 7.5 mark is drawn in merely to highlight the difference between the two sets of results. Note that the two lowest ranking chicks in each of the light-exposed groups gained fewer entries to the food dish than did their equivalents in the dark-incubated groups. Data adapted from Rogers and Workman (1989).

embryos communicate with the hen (Guyomarc'h, 1966; Tuculescu and Griswold, 1983) and in response to hearing the embryo's distress calls the hen rises to turn the eggs. At this moment light may reach the eggs. Therefore, the frequency of such embryo–hen interactions should determine the amount of light received by the embryo. As measured by Tuculescu and Griswold (1983), between 90 and 100 distress calls were emitted by the embryos in a clutch during a 30 min interval of recording at three hours before hatching (see Figure 2 in Tuculescu and Griswold, 1983). Of course, the hen would not respond to every individual call, but, as only two hours of light exposure are sufficient to establish lateralization of brain structure and function, these results indicate that over the last days before hatching chick embryos incubated by the hen might receive sufficient exposure to light for lateraliza-tion to be established. Moreover, the hen may leave the nest for longer periods of time to search for food, and this would also allow light exposure of the eggs. Unfortunately the temporal patterning of light exposure to which eggs are more likely to be exposed during natural incubation has not yet been investigated as a factor influencing the development of brain asymmetry.

Other Factors that May Influence Development of the Embryo

The effects of auditory and visual stimulation on the development of the embryo and their consequent influence on behaviour after hatching have been the only forms of sensory stimulation studied in any detail. There are, however, other forms of sensory stimulation of the embryo that may also influence its development.

Olfactory stimulation

As mentioned in Chapter 1, the olfactory system of the chick embryo becomes functional by day E20 at the latest (Tolhurst and Vince, 1976) and thus the embryo may potentially imprint on the odour of its surroundings. That is, prior to hatching an attachment may form to the odour of the nest and/or the hen.

It is also possible that olfactory stimulation interacts with input to any of the other sensory modalities, as these are all functional at the stage when the olfactory system becomes functional.

Influence of atmospheric pressure

Bateson (1974) has reported data showing an association between atmos-pheric pressure and imprinting on day 1 after hatching. The imprinting behaviour was measured as approach to a familiar stimulus (Chapter 3). There was a highly significant negative correlation between approach behaviour and the barometric pressure which had occurred on day 12 of the incubation period. In addition, the median length of the incubation period was positively correlated with the pressure on this day. Hatchability was not affected.

Bateson suggested that the atmospheric pressure may exert its effect by altering oxygen and carbon dioxide concentrations in the blood at a critical stage of nervous system development. Indeed, the partial pressure of oxygen and weight-specific rate of oxygen consumption in chick embryos increases transiently midway through the incubation period, and these changes correlate with changes in antioxidant enzymes in the brain and blood (Wilson *et al.*, 1992). These antioxidant changes are related to oxygen supply, and they may reflect proliferation and differentiation of certain cell types. For example, beginning on day E12 the level of glutathione peroxidase in the brain increases along with an increase in the number of glial cells (Tsai *et al.*, 1981). Thus, changes in atmospheric pressure may indirectly influence brain development, and so have behavioural consequences after hatching.

Changes in atmospheric pressure may also affect the blood pressure of the embryo as a result of altering the respiratory milieu. Mean arterial blood pressure, which may influence nutrient supplies to the developing brain as well as other organs and tissues, normally rises at an exponential rate from day E3 to E10 and then continues at a lesser rate of increase until hatching (Girard, 1973). Heart rate also increases with embryonic age but continues to do so up until day E15 (Soliman and Huston, 1972). Alterations in this pattern may potentially have consequences on brain function, although this is not yet known.

Intersensory Stimulation During Development

The sequential development of the various sensory systems means that the initial sensory input in certain modalities occurs in the absence of input in another modality. For example, since the auditory system matures before the visual system, the embryo receives and processes auditory information before it does so for visual information. That is, the initial auditory inputs do not compete with visual inputs. This orderly sequence of sensory stimulation appears to be essential for processing of intersensory information after hatching (Banker and Lickliter, 1993; Lickliter, 1993). If the developing embryo is stimulated by patterned visual inputs just prior to hatching, social responses after hatching are altered. This has been demonstrated by Lickliter (1990) using bobwhite quail. Quail chicks hatched from eggs which have been in a darkened incubator rely at first on auditory cues only to identify their own species (as appears to be the case for domestic chicks incubated in darkened incubators, Gottlieb and Simner 1969), but those that have received temporarily patterned visual stimulation before hatching require both auditory and visual cues to make the same identification.

Lickliter (1990) assessed the intersensory functioning of the bobwhite quail chicks by testing them at 24 or 48 h after hatching for approach to either their species-specific maternal call or the maternal call of a domestic chicken, and then with another choice between a stuffed bobwhite quail hen coupled with the bobwhite quail maternal call and a stuffed hen of another species of quail coupled with the chicken maternal call (Lickliter and Virkar, 1989). The controls had the eggshell and membranes removed at the air sac end of the egg but did not receive extra visual stimulation, whereas the test group were treated similarly and exposed to patterned visual stimulation, a light flashing at three cycles per second. In the test using auditory cues alone the control chicks chose to approach their species-specific maternal call, whereas the chicks that had received the visual stimulation showed no preference for either auditory stimulus. In the choice test with stimuli presenting visual plus auditory cues, however, both the test and control chicks preferentially approached the bobwhite quail emitting their species-specific maternal call. In other words, the chicks which had been incubated in darkness (the controls) could choose on the basis of auditory cues alone, whereas those which had received visual stimulation prior to hatching responded to their species-typical characteristics only when presented with both visual and auditory cues.

Lickliter (1990) interprets these results as showing that 'premature' sensory stimulation with patterned light accelerates the development of intersensory function. In addition, the light-stimulated embryos may develop lateralization of the visual pathways, as shown for the domestic chicken, and lateralized functioning of visual information in the forebrain (see p. 63). The dark-incubated controls would, apparently, lack lateralization for visual processing, although they may well have lateralization for auditory processing. Whether it is the combined lateralization of auditory and visual processing in the light-exposed chicks which leads to the requirement for intersensory integration of visual and auditory information has yet to be determined.

Lickliter argues that the visual stimulation provided to his test group is premature (i.e. abnormal). Admittedly, he removed the shell and membranes at the air sac end of the egg so that the right eye of the embryo received more light simulation than usually occurs, but light is able to pass through the shell and membranes of the intact egg. Although the eggs may not be exposed to light flashing at exactly 3 cycles per second in the natural environment, the experimental procedures did not investigate whether other forms of light stimulation, even continuous light, were similarly effective. In the natural environment light stimulation of the eggs might well occur prior to hatching, even as flashing light when the hen stands up and sits down (see p. 66). Thus, contrary to Lickliter's statements, the group exposed to the flashing light may be closer to the normal condition. If so, Lickliter's experiments involve altering the normal course of development by depriving the 'controls' of light experience which might normally occur just prior to hatching, rather than providing premature light experience to the 'test' animals.

Nevertheless, since changing the sequence of sensory stimulation affects intersensory functions after hatching, this research shows that it is important that the different sensory systems become functional at different times during embryonic development. Thus, as shown by Lickliter (1990), competitive relations between the maturing sensory modalities are limited or permitted by the timing of the development of each sensory system. Although tactile inputs must interact with the initial processing of auditory information, because the former matures some seven days before the latter, auditory processing and memories are established well before visual inputs can be fully attended to by the developing embryo.

Even within one sensory system there may be a sequence of development that allows the embryo to receive certain forms of sensory stimulation and not others. For example, the visual system of the embryo becomes functional prior to hatching at a stage when the eggshell and membranes prevent the eyes from receiving patterned visual stimulation. Thus, visual processing begins with information about changes in light intensity and wavelength (albeit limited by the shell) and only later is patterned information processed. This sequence appears to be important for the complete development of visual functioning.

So far there have been relatively few studies of such interactions between stimulation in different sensory modalities and the process of development. The research by Lickliter and colleagues demonstrates that the sequence of sensory development leads to functional priorities of one sensory system over another at different stages of development, and that stimulation in one or more of these systems interacts with the others to qualitatively influence the mode of neural processing adopted after hatching (Lickliter and Virkar, 1989). Stimulation in one sensory system can also accelerate the maturation of another sensory system: Lickliter and Stoumbos (1991) were able to accelerate species-typical visual responsiveness by exposing bobwhite quail embryos to increased levels of their species-typical calls. As originally pointed out by Schneirla (1966) and Kuo (1932a,b) there is an intimate interaction between maturation of the nervous system and the experience which it receives.

Lickliter's experiments have investigated interactions between only two of the sensory systems. Consideration of interactions between all of the sensory systems is now timely, but bound to be extremely complex. These cross-modal interactions are now being actively studied in mammalian species. For example, deprivation of vision from birth can be compensated for by enhanced dependency on tactile and auditory senses. In cats and mice deprived of vision the facial vibrissae undergo supernormal growth and the neural representation of this system in the brain, the barrelfield, expands accordingly

(Rauschecker *et al.*, 1992; Bronchti *et al.*, 1992; Diamond *et al.*, 1993). Also in cats, it has been found that early blindness leads to sharpening of auditory spatial tuning of cortical neurons (Korte and Rauschecker, 1993). In these species we are dealing with postnatal effects over weeks or months of visual deprivation, but the marked changes in sensory processing which result are similar to those apparent after much shorter periods of deprivation or experience in the avian embryo, and both may presumably rely on similar neural mechanisms.

The neural plastic changes which lead to compensatory changes in the sensory systems of mammals are now known to depend on the presence of a particular subtype of glutamate receptor, the NMDA receptor (Rauschecker, 1991), but so far this has not been studied in the developing avian embryo. NMDA receptors, however, are known to be involved in the neural plastic changes which follow imprinting in chicks, as will be discussed in Chapter 4.

Concluding Remarks

Most of the studies which have investigated the influence of environmental factors on the development of brain and behaviour in the chick embryo have involved either deprivation of stimulation of one sensory modality or, apparently, excess stimulation in one modality. Under natural conditions of incubation such unitary changes in only one sensory modality are likely to be rare.

Of course, one of the advantages of using chick embryos to study behavioural embryology is the accessibility of the egg and embryo to experimental manipulation in the laboratory setting, but not one of these experimental procedures which involves the use of artificial incubation seems to mimic closely natural incubation by the hen. There is now clear evidence of bidirectional communication between the embryos and the hen, at least in the later stages of incubation (Tuculescu and Griswold, 1983), and of communication between the embryos themselves (Vince, 1964; Impekoven and Gold, 1973). This communication may be instrumental in causing simultaneous exposure of the embryo to sensory stimulation in a number of modalities. Thus, distress calls by the embryo elicit egg shifting and vocalization by the hen. As the hen stands up to move the eggs, the entire act causes simultaneous stimulation of the vestibular, tactile, proprioception, visual and auditory systems of the embryo. Moreover, the initiation of this sequence of events by the embryo guarantees that the embryo will be awake at the time the stimulation occurs.

Therefore, unlike the artificial paradigms used in the laboratory, natural incubation appears to maximize multiple sensory inputs at the same time. Judging from the findings of Lickliter (1990), this must result in the development of optimal intersensory integration by the time the chicks have hatched, and presumably a requirement for processing multiple cues from more than one sensory modality, together with an ability for performing intermodality generalizations.

We must also be reminded that each embryo provides itself with stimulation via its own vocalizations and movements within the egg. Thus, both the structural and functional development of the nervous system is influenced by a wide range of interacting factors from the environments internal and external to the egg.

3

EARLY LEARNING AFTER HATCHING

Summary

- Shortly after hatching the young chick imprints on the visual characteristics of the hen. This is known as filial imprinting. Sexual imprinting occurs later in the chick's life.
- Over the first one or two days of posthatching life the chick develops a predisposition to approach a complex visual stimulus, such as the hen.
- Imprinting to the maternal call may occur prior to hatching, and auditory and visual associations may form after hatching as the chick continues to learn about the hen. Auditory cues direct the chick's attention to the visual imprinting stimulus, leading it to learn multiple parameters of the visual imprinting stimulus. Olfactory imprinting may also occur, although there is far less information on this form of imprinting.
- Imprinting occurs to peers as well as the hen. Brood mates are recognized from strangers, and the hen recognizes her own chicks.
- Fear behaviours and responses to alarm calling by the hen develop over the first few days posthatching.
- Learning to feed takes place over the first week or two posthatching, and it is influenced by social factors which facilitate pecking and establish preference for particular types of food. Beak-trimming alters food intake and food preferences.
- Visual mechanisms used in feeding are discussed, as also are tactile, olfactory and taste aspects. Many of the visual behaviours associated with feeding are known to be lateralized.
- Very young chicks form social hierarchies based on access to food and other resources or on attack leaping behaviour. Later the social hierarchy can be measured in terms of aggressive pecking.
- The time spent sleeping declines consistently as the chick ages, but there are peaks of sleeping at particular ages.

Introduction

The domestic chick hatches not only with a brain that is ready to undergo learning critical for survival, but also with one that has already established some memories and has been shaped by the environmental context of incubation. In particular, the nearly hatched chick is likely to have learnt at least some of the characteristics of the hen's calls and possibly to have established a familiarity with the olfactory environment of the nest and hen. Additionally, non-patterned visual stimulation has primed its visual system and influenced the visual pathways to develop asymmetrically. After hatching, the chick is exposed to patterned visual stimulation and must learn the visual characteristics of the hen and food objects. For at least the first three days after hatching there are still sufficient nutrients available from the yolk sac to make the ingestion of food less a matter of survival (Freeman, 1965). Although the newly hatched chick immediately pecks actively at small three-dimensional objects (Dawkins, 1968), it usually does so with a closed beak. This exploratory pecking will lead on to pecking with ingestion of grains and so to learning to discriminate food from inedible objects, as will be discussed later in this chapter, but first a powerful form of learning by the chick known as imprinting will be discussed.

Filial Imprinting

The most important learning that occurs in the period immediately posthatching is learning to recognize the hen by a process termed imprinting. At first, the hatched chicks sleep a lot and huddle in the nest next to the hen (Bateson, 1987); they gradually become more active over the first 12–16 h after hatching (Hess, 1959a). The increase in activity occurs along with developing motor abilities, manifested by the adoption of an increasingly upright posture as they stand up on their legs and obtain improved motor coordination. Vince and Chinn (1972) assessed the development of motor abilities in domestic chicks stimulated in various ways. They found that chicks hatched in a brooder with other chicks are all able to stand upright by 7.5 h after hatching, and motor coordination for walking improves after this. Various degrees of social isolation after hatching delay this development, but they are relevant to the natural situation only to a limited extent, as an indicator of the time at which the slowest developing chick may be expected to have reached this particular stage of development. This appears to be by the end of the first day posthatching.

Once locomotor ability has developed sufficiently for the young chick to leave the nest, it approaches conspicuous visual stimuli, and visual imprinting occurs. In particular laboratory settings the optimal period for visual imprinting is at around 12–24 h posthatching (Ramsay and Hess, 1954;

Sluckin, 1966), but this period may vary according to sensory and social experience (Bateson, 1966). For ducklings reared in the laboratory, the optimal time for visual imprinting is around 12–20 h posthatching (Hess, 1959a) and it is known that the optimal time for visual imprinting follows that for auditory imprinting (Gottlieb and Klopfer, 1962), a result which is not surprising given the earlier development of the auditory system. Although it is possible to visually imprint a chick on the first day after hatching, under natural conditions of incubation visual imprinting may not begin to occur until after the first day of life posthatching (Guyomarc'h, 1975; Bateson, 1987).

For the first few days after hatching the chick shows no fear of conspicuous visual stimuli and, if such a stimulus is moving, the chick will follow it (Bateson, 1966; Sluckin, 1966). During the period of exposure to the visual stimulus, the chick rapidly forms a memory of its characteristics and thereafter it exhibits distress behaviour when separated from that particular stimulus. As the learning that occurs proceeds rapidly, without any obvious reward, and leaves a stronger memory trace than virtually any other learning that the chick may do in later life, it has been distinguished by being called imprinting (Lorenz, 1935). More recently it has become clear that imprinting is associated with the same biological processes as other forms of learning (Chapter 4), and so it may differ only in the context in which it occurs in early life and its impact on later behaviour (Bateson, 1990).

Imprinting on the hen has been termed filial imprinting, in contrast to sexual imprinting. Filial imprinting has an immediate consequence in that it leads to following of the hen by the chicks. Sexual imprinting occurs somewhat later in life and leads to choice of conspecifics as sexual partners (Bateson, 1979a; Vidal, 1980). Sexual imprinting is affected by the chicken's experience with other birds, particularly the hen and the siblings. The effects of these experiences are not manifest until later in life when sexual maturity is reached. Originally Lorenz (1935) considered imprinting to be a single process which led to following the hen in early life and determined sexual preferences in later life. We now know that filial imprinting is distinct from sexual imprinting (see later for discussion of sexual imprinting).

After filial imprinting the clutch of chicks follows the hen when she moves and is thus protected from predators. In addition, close proximity to the hen enhances learning about the hen's behaviour by the chicks, for example, learning to peck at food objects (see pp. 99–101).

Filial imprinting is not just a following response; the term refers also to the acquisition of a social preference, since the social behaviour of the chick becomes directed toward the imprinting object irrespective of whether it is moving or not (Salzen, 1966; Bolhuis, 1991). The chick remains close to the imprinting object, often huddling against it, frequently pecks gently at it and shows contentment behaviour, including pecking at food particles and making pleasure calls (twitters).

Visual imprinting

Filial imprinting to a visual stimulus has been measured in a number of ways, all designed to assess either the chick's approach to or following of the imprinting stimulus, often relative to a different novel object (Sluckin, 1966; Horn, 1985), or the chick's reduction in distress calling (peeping) in the presence of the familiar stimulus (Boakes and Panter, 1985; de Vos and Bolhuis, 1990). Figure 3.1 represents an apparatus used frequently to test imprinting in precocial species. The chick's following response is scored in terms of its distance from the moving imprinting object over a fixed time interval of around 20 min. The imprinting stimulus may be moved in a random manner in an open field instead of being rotated in the circular runway, or it may be moved back and forth in a straight runway (Jaynes, 1956). The movement of the imprinting stimulus may also be interrupted by pauses.

Imprinting can also be measured in terms of a preference for the imprinting stimulus over another unfamiliar stimulus. Figure 3.2 shows a chick given such a choice, its approach or avoidance of the imprinting stimulus being assessed by rotations of the running wheel. This method of imprinting, which has proved most useful in study of the neurobiological basis of learning and memory, was developed by Bateson and Jaeckel (1976) and Bateson and Wainwright (1972), and it has been used extensively by Horn, Bateson and co-workers at Cambridge University to investigate the neurochemical correlates of imprinting memory formation (Horn, 1985; Chapter

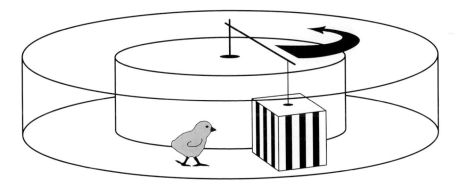

Fig. 3.1. An apparatus which has been used to test the following response of young chicks or ducklings. The visual stimulus (in this case, a box with striped walls) is moved slowly in a circular runway. The model may also emit a sound, such as the hen's cluck call. Over a testing period of, for example, 20 min the following response of the chick is scored in terms of its distance from the stimulus. Other responses such as avoidance of the stimulus (including freezing), latency to commence following and pleasure and distress calls can be scored. The diagram depicts the apparatus described by Bateson (1964a) and Gottlieb (1965a); it follows the original design of Hess (1959a).

Fig. 3.2. Imprinting preference can be measured by placing the chick in a running wheel attached to a device which counts the number of revolutions in either direction, and presenting the chick with a choice of two, rotating visual stimuli placed on either side of the wheel at right angles to its axis. The apparatus is an adaptation of that originally designed by Bateson and Wainwright (1972). It is the one commonly used in studies of memory formation following imprinting by Horn and colleagues (Horn, 1985) and in my laboratory (Johnston *et al.*, 1993). The chick is first imprinted by exposing it to either the box or the hen. It is then tested for its percentage preference with a choice between the imprinting stimulus and the unfamiliar stimulus. A preference score is calculated by dividing the number of revolutions of the wheel made as the chick attempts to approach the imprinting stimulus by the total of the number of revolutions in each direction. This calculation controls for differences in activity between chicks. This particular chick has been imprinted on the hen and here it is attempting to approach that stimulus. The rotating box is positioned in the top right corner of the photograph.

4). With slight modifications, the method has been used in other laboratories. In the training trial the chick, previously held in the dark except perhaps for a brief period of exposure to diffuse light (priming) which enhances the imprinting (Bateson and Seaburne-May, 1973), is placed in the running wheel and exposed to the imprinting stimulus. The latter is presented some 50 cm from the wheel. The imprinting stimulus might be a rotating, coloured box (e.g. with two opposite faces painted red and the other two black) or a rotating stuffed hen. The object is rotated with intervening pauses in its motion. The exposure time varies, but is often around 140 min. The chick is then returned to the dark until testing occurs after an optional delay period. Testing involves giving the chick a choice between the imprinting stimulus and a novel stimulus. The latter might be a differently coloured box or differently coloured hen. The two stimuli are placed on either side of the running wheel, into which the chick is placed. For a five minute period the number of rotations made by the wheel in the clockwise and anticlockwise directions is counted. These rotations are generated by the chick as it attempts to approach either the imprinting stimulus or the novel stimulus. A percentage preference score is calculated as the number of rotations toward the imprinting stimulus divided by the total number of rotations in either direction. This score excludes any contribution that could be made to the estimated preference by differences in activity between individuals, which can be a problem with other methods of testing. A high percentage (usually in the region of 70–80%) indicates a preference to approach the training stimulus and therefore that the chick has imprinted. An unimprinted chick shows no preference, scoring around 50%. Alternatively, the imprinting stimulus (A) and the unfamiliar stimulus (B) are presented sequentially one at a time, usually in the order ABBA for 2–4 min each, and the percentage imprinting calculated accordingly.

The chick does not have to follow or attempt to approach the stimulus to imprint. It will also imprint on an object suspended in the home cage (Salzen and Meyer, 1968; Salzen *et al.*, 1975). Strong imprinting occurs with 1–2 h of exposure to a novel visual stimulus, but, as Salzen and Sluckin (1959) found, even as little as a few minutes of exposure on day 1 to a moving box leads to some degree of imprinting. As might be expected, the required duration of exposure depends on the nature of the visual stimulus. Moving stimuli are preferred over stationary ones (Ten Cate, 1989a), red stimuli over yellow ones (Kovach, 1971; Bateson and Jaeckel, 1976; Bateson, 1983a), and ball-shaped objects over rectangular box-shaped ones (Salzen and Meyer, 1968).

Attention to multiple parameters of the visual imprinting stimulus

Not only does imprinting on the shape of an imprinting object depend on its colour (see pp. 82 and 83), attention to more than one colour of a multicoloured imprinting stimulus depends on the preference for one of those colours. To illustrate this, van Kampen *et al.* (1994) imprinted jungle fowl chicks on objects with two colours and then tested them for preferences with one of the colours changed. Chicks imprinted on an object coloured red, a preferred colour, and yellow noticed when the yellow elements of the stimulus were changed to white, whereas chicks imprinted on a black, less preferred colour, and yellow stimulus did not react when the yellow elements were changed to white. That is, attention to the yellow element of the imprinting stimulus occurred only when the other colour was red. Rather than overshadowing learning about the yellow colour, the preferred red colour enhanced learning about it. This means that, if the natural mother has at least some of her plumage the preferred colour, the chicks may attend to and remember all of the colours of her plumage. By extrapolation, this would be a mechanism for ensuring attention to detail for recognition of individuals.

The position of the imprinting stimulus is also a feature that is learned by the chick. A chick trained on an imprinting object placed in a quadrant at the back of its cage for seven days prefers to remain in that quadrant when the imprinting object is no longer present and, when the imprinting object is placed in another quadrant, the chick shows more distress calling than when the object is in the training location (van Kampen and de Vos, 1992). However, the distress calling persists only during the first 5 min after placement of the imprinting stimulus in a novel location: thereafter, the chick generalizes to accept the imprinting stimulus in a new position. Features of the imprinting stimulus other than position (i.e. colour and shape) are held more stably, and so the chick will follow the object as it moves. Thus, although position is a feature to which the chick attends, it does not lead to formation of a stable preference for position, and nor is the strength of the position memory dependent on the effectiveness of the object as an imprinting stimulus (van Kampen and de Vos, 1992). It should be noted that position cues may have a negligible role when the chick imprints naturally on a hen, since the hen will be moving and certainly never seen only in a single location, as was the imprinting object in these experiments.

Imprinting on the visual environment

The chicks may imprint on aspects of the environment other than con-spicuous single objects. Laboratory experiments have shown that they can form a preference for the walls of their home cage (Bateson, 1964a), suggesting that in the natural environment some early form of learning akin to imprinting may occur to features of the environment surrounding the nest. Bateson raised the chicks for three days in a cage with striped patterns on the walls and then tested with a choice between two boxes, one painted with the same pattern as the walls of the home cage and the other with an unfamiliar pattern. The chicks preferred to approach the box painted like the walls of the home cage. Figure 3.1 illustrates the apparatus used by Bateson (1964a) to test the chick's preferences. In the experimental design, however, the chicks were not raised with simultaneous exposure to a moving visual stimulus and the walls of the cage, and so it is not known whether learning about patterns in the environment would occur in the presence of a preferred imprinting stimulus, such as the hen. If chicks can learn about, or even imprint on, features in the surrounding environment during the first day after hatching, and this has not yet been tested, it is possible that prior to imprinting on the hen, and while still in the nest, they begin to form an association with their surrounding environment. They might do this by looking out from the nest, although there are no data documenting whether this behaviour occurs, or not. Nevertheless, Broom (1969) has shown that chicks do learn many of the features of their environment between days 2 and 3, as indicated by increasing responsiveness to changing that environment by illuminating a small light bulb. Visual features of the nest itself and of nearby conspicuous visual objects may be learnt prior to following the hen and imprinting on her. Observations by Guyomarc'h (1975) suggest that the chicks see little, if anything, of the hen in her entirety until about day 2 posthatching. On day 2 the chicks begin to emerge from under the hen, but only for short periods of time. By days 3 and 4 the chicks spend time beside the hen, and then, beginning on day 5, they actively follow the hen for much of the time. By this age the chicks have already imprinted on the hen.

The emergent predisposition

Furthermore, there is a predisposition for the chick to approach a hen-like stimulus (e.g. a rotating stuffed hen as opposed to a rotating coloured box). It is the head and neck region of the hen which particularly attracts the chick; Johnson and Horn (1988) demonstrated that chicks do not prefer a model of a complete hen over one with only the head and neck region. In fact, there is no species-specificity in the predisposition preference, because the chick shows an equivalent preference for stuffed models of other species (Johnson and Horn, 1988).

The predisposition develops over the first one to two days of life (Johnson *et al.*, 1992) so that, although on the first day after hatching a chick can be imprinted on a box or a stuffed hen with equal ease, by the next day the hen is a preferred stimulus. This preference for the hen appears in visually experienced chicks which have not seen a hen before. It can also emerge irrespective of prior imprinting learning, as Johnson *et al.* (1985) have demonstrated. They imprinted chicks on either a rotating red box or a rotating stuffed jungle fowl and tested them with a choice between these two stimuli at either 2 or 24 h later, using an apparatus similar to that pictured in Fig. 3.2. When tested 2 h later, each chick preferred the stimulus to which it had been exposed previously, either the box or the fowl. By 24 h, however, the situation had changed. Those exposed to the fowl showed an increased preference for that stimulus, whereas those exposed to the box had lost their preference and performed at chance level. That is, they too were showing an increased preference for the fowl in the choice test. In fact, the predisposition for the fowl develops even with the non-specific experience of simply being placed in the running wheel, used during the choice tests (Fig. 3.2), and exposed to dim light in the featureless testing chamber (Horn, 1985), although it appears earlier in chicks exposed to a complex visual pattern compared to those held in darkness (Bolhuis *et al.*, 1985). At 24 h after this exposure, the chicks showed a preference for the fowl over the box. Controls held in the dark had no such preference. There is a sensitive period for development of a predisposition in response to being placed in the running wheel. Chicks placed in the running wheel at 24 and 36 h of posthatching life develop the predisposition 24 h later, but those treated similarly at 12, 42 or 48 h do not (Johnson *et al.*, 1989). Therefore, under these conditions, the predisposition is evident sometime after 12 h and before 42 h. Of course, this time period may vary under other rearing conditions, but a sensitive period is still likely to be present.

Hormonal condition also affects the emergence of the predisposition. Bolhuis *et al.* (1986) found that chicks treated with testosterone (0.5–5 mg of testosterone oenanthate injected between 3–8 h posthatching) imprinted more strongly on a stuffed jungle fowl, but not on a red box. In fact, preference scores for fowl-trained, but not box-trained, chicks were found to correlate with plasma testosterone concentrations. It remains to be seen whether other sex steroid hormones have similar effects and thus whether sex differences exist for the predisposition.

A more recent series of tests by Johnson *et al.* (1992) has shown that the shift away from a preference for the red box by chicks imprinted on the red box is not simply due to more rapid forgetting of that particular stimulus. When given a choice between the red box and a striped blue box at 24 h after training, these chicks showed no decline in their preference for the red box. Thus, the developing preference for the hen is not merely a consequence of losing a preference for the red box, but rather an independent phenomenon, likely to be carried out in a neural system separate from that involved more directly in imprinting (Johnson *et al.*, 1992). The two systems apparently compete for control of approach behaviour.

Under natural conditions the hatched chicks visually and auditorily imprint on the live hen, and the environmental context is designed to ensure that this will occur. Although initially the chicks might be attracted by a wide range of conspicuous objects, particularly if those objects are moving, the delay in the development of their locomotor ability would ensure that the predisposition to approach stimuli with characteristics of conspecifics (i.e. the hen) has emerged by the time they leave the nest to approach and follow moving objects. Thus, it is likely that the young chicks do not begin to imprint visually until the predisposition has emerged to 'guide' their attention toward the hen (Bolhuis *et al.*, 1985). Furthermore, the hen is likely to be the most attractive moving visual stimulus in their vicinity. Thus, once the chicks have been exposed to the hen and they have imprinted on her individual characteristics, they may be less likely to be attracted by other stimuli.

In fact, Boakes and Panter (1985) found that prior exposure of chicks on days 1 and 2 to a hen suppressed secondary imprinting on a moving cup, whereas prior exposure to a moving windmill did not. The hen appeared to be the most potent imprinting stimulus in terms of its ability to suppress secondary imprinting on another stimulus. Similarly, Bolhuis and Trooster (1988) found that when a chick is imprinted first on a stuffed hen it will not imprint on a red box but, when the first imprinting is on the box, the preference is altered by secondary exposure to a hen. Contrary to these results, a more recent study by de Vos and van Kampen (1993) has shown that imprinting on a naturalistic stimulus (a live chick) postpones, but does not block, secondary imprinting on an artificial stimulus (a yellow cylinder). If Boakes and Panter (1985) had tested after a longer delay period following training, they may have found the same result.

Thus, contrary to the original concept of Lorenz (1935), imprinting on the hen is reversible in the sense that secondary imprinting can occur, but is not as easily reversible as imprinting to an artificial object. It appears that, when the imprinting stimulus is relevant for the predisposition (a live or stuffed hen, or chick), the two interacting mechanisms of predisposition and imprinting are engaged and a relatively more stable preference forms (Bolhuis, 1991). Even though the chick will imprint on a second stimulus, it does not forget the first and, in that sense at least, the imprinting is not reversible (Cook, S.E., 1993).

It cannot be insignificant that the predisposition to approach the hen is timed to emerge when the chick's motor abilities permit it to leave the nest and follow (Bateson, 1987). One might question why the predisposition is not present from the time of hatching. It is possible that the sensory systems require maturation after hatching, possibly in an experience-dependent manner. Indeed, stimulation with spatially patterned visual stimuli cannot occur until hatching because the eggshell and membranes prevent this information from reaching the embryo, and prior processing of such information may be required before the predisposition can develop. Consistent with this possibility, a preference for the hen does not develop in chicks housed individually in a dark incubator (Johnson *et al.*, 1992).

Auditory imprinting

Auditory learning about the hen's vocalization may also have occurred prior to hatching (Chapter 2) and it is likely to continue in association with visual imprinting after hatching (Fält, 1981). Thus, prehatching auditory learning, possibly superimposed on a predisposition for the species-specific maternal call (Gottlieb, 1965b), may act to focus the newly hatched chick's attention on the hen: a visual model that emits sounds is preferred over a silent visual model (Smith and Bird, 1963, 1964). Thereafter, auditory and visual associations may form as the chick continues to learn about the imprinting stimulus, the hen. Kent (1987) has shown that chicks aged 4 or 8 days can discriminate their own hen's maternal cluck call from that of other hens. Following exposure to a clucking sound, the chicks will follow a speaker emitting that sound and emit distress calls while they do so (Graves, 1973). The distress calls emitted in this context are explained by Graves (1973) as having a role in contact-seeking by the apparently 'lost' chick seeking its mother. Auditory imprinting, however, does not seem to be as important as visual imprinting (Evans, 1982; Bolhuis and van Kampen, 1992), at least in the sense that it does not appear to persist for as long. After four hours' separation from the hen on day 4, the chicks no longer discriminate the call of their own hen from that of another hen (Kent, 1987).

Auditory cues may serve to direct the young bird's attention to the visual stimulus, but the chick apparently does not form a lasting association with the sound itself (Gottlieb and Simner, 1969; Klopfer and Gottlieb, 1962). In fact, chicks may not develop a consistent preference for a sound alone, but rather for a sound in association with a visual stimulus. For example, van Kampen and Bolhuis (1991) found no evidence for a significant preference for a pure tone in chicks previously exposed to that tone, whereas those exposed to the same tone together with a rotating red box did develop a preference for their training stimulus. A similar result was found in an earlier experiment which used a parental call paired with a visual stimulus (a moving pendulum) or the call alone (Cowan, 1974). When the chicks were tested with a choice between the familiar call and a novel parental call, only those exposed to the combined auditory and visual stimulus showed a preference for the familiar call. The chicks exposed to the auditory stimulus alone had not imprinted. Bolhuis and van Kampen (1992) have suggested that the visual stimulus may act as an unconditioned stimulus for learning in the Pavlovian design (see also Bolhuis *et al.*, 1990).

Furthermore, the presence of a maternal call together with a visual stimulus has been shown to enhance visual imprinting tested in the absence of the call (van Kampen and de Vos, 1991; van Kampen and Bolhuis, 1993). In other words, the combination of auditory and visual cues leads to enhanced visual imprinting. Taken together, the results of van Kampen and Bolhuis (1991, 1993) show that combining auditory and visual stimuli enhances learning in both modalities separately. That is, rather than one salient feature (visual or auditory) overshadowing or blocking the other, both features are learned and can be responded to separately. This lack of overshadowing by a salient feature also occurs with a single modality. For example, chicks imprinted to a red object (known to be a preferred colour for imprinting; Schaefer and Hess, 1959) still respond to changes in its shape (van Kampen and de Vos, 1991; van Kampen, 1993b). Imprinting on a yellow object, a less preferred colour, does not, however, lead to attention to its shape. Thus, the more effective an object is for imprinting, the more the chick learns about each of its separate features. These results suggest that imprinting is not identical to associative learning, although many other aspects of imprinting can be explained by associative-learning theory (Bateson, 1990; van Kampen, 1993a).

Olfactory imprinting

In Chapters 1 and 2 it was noted that the embryo is capable of responding to olfactory stimuli, and therefore olfactory memories about the nest and hen may be laid down prior to hatching. These may provide one of the influences which serve initially to keep the newly hatched chicks in the nest huddled next to the hen. Along with the auditory associations made before hatching, the olfactory learning may also function to focus visual attention on the hen. If so, however, this would probably occur only over a short range of distance, in which the odours would diffuse. After hatching the chicks might also undergo further olfactory learning, possibly imprinting. In fact, experiments by Vallortigara and Andrew (1994) and Burne and Rogers (1994a,b) have demonstrated that it is possible to imprint chicks on odours to which they are exposed in early posthatching life. In both cases, the chicks were housed individually in cages each with a container suspended at the same height as the chick. The containers were filled with cotton wool either unscented or scented with clove oil, orange oil or amyl acetate (Vallortigara and Andrew, 1994), or garlic or nesting material (Burne and Rogers, 1994a,b). At either day 3 or 4 the chicks were tested in a runway with an unscented container suspended at one end and their familiar-scented container at the other end. Both stimuli had identical visual characteristics. The chicks were found to spend a significantly greater amount of the total testing time in the vicinity of the container containing their familiar odour, except in the case of those raised with the garlic odour. The latter chicks did not develop a preference for the garlic odour until they had been raised in its presence for around two weeks (Burne and Rogers, 1994a). In each case, it was shown that the chicks were definitely choosing the familiar stimulus on the basis of its odour, and not on fine visual cues which may have been learnt, by testing the chicks with stimuli of changed shape. The chicks still chose their familiar odour. They had been imprinted on the odour with which they had been reared.

Recognition of the chicks by the hen

Although most research to date has focused on recognition of the hen by the chicks, the hen must also recognize her own brood of chicks. Kent (1992) fostered groups of either brown or black chicks to hens and then tested them in an open field with a choice between their brood and an alien brood. The hen chose her own brood when it was matched with a brood of a different colour, but she chose an alien brood of the same colour as her own when asked to make a choice between the latter and another alien brood of an unfamiliar colour. Thus, the hen recognizes her brood on the basis of colour. Kent (1992) generalized these results to say that they rule out the possibility that auditory cues are used, but before this conclusion can be reached the hen should be tested with a choice between her own brood with changed colour and an alien brood of a different colour. Also, more detailed analysis of the hen's behaviour to the brood when she has chosen an alien brood on the basis of colour might reveal that, although recognition on the basis of colour is primary, auditory cues also have a role in her recognition of the chicks, and perhaps olfactory cues do too.

Peer attachment and imprinting

Chicks also learn to recognize their brood mates and then to follow them, which means the group (hen plus chicks) stays together (Salzen and Cornell, 1969). In fact, they probably become imprinted on their siblings using colour as an important cue. Kilham *et al.* (1968) found that socially reared chicks chose to approach a chick of the same colour (yellow or blue) as its brood mates. Moreover, brood mates are recognized as individuals. Male chicks raised with a companion chick and then tested in a runway with a choice of that companion and a strange chick avoid the stranger (Vallortigara and Andrew, 1991). A cage mate is also more effective in reducing fear in the open field than is a stranger, but only in chicks housed in pairs, not in isolation (Jones, 1984). In fact, Zajonc *et al.* (1975) have reported that even one-day-old domestic chicks can discriminate between familiar and unfamiliar members of their own species. This was determined by scoring social pecking of companions or strangers in paired encounters.

The chick must be able to recognize familiar siblings even though their plumage and other characteristics change with age. Bateson (1979a,b, 1990) has explained that this may be made possible by the imprinted chick's preference for a slightly novel stimulus compared to the imprinting stimulus itself. Thus, they may be able continually to update the representation of a sibling. This preference for slight novelty might also explain why a chick imprinted on a stuffed hen will work to obtain novel views of the stuffed hen (Bateson, 1979b). In fact, by this interest in slight novelty the chick may build up a complex representation that will allow it to recognize the hen at all angles of viewing. The same is likely to occur for the representations of each of the chick's siblings.

Not only do chicks recognize their peers but they appear to interpret the social interactions between chicks. At least this was the conclusion reached by Regolin *et al.* (1994), who tested chicks on the traditional problem of going around a barrier using other chicks as an incentive goal. The chick being tested was more motivated to approach a pair of interacting, socially reared chicks than an isolated chick or a pair of chicks not interacting socially. It would seem that the experience of social rearing teaches the chick how to recognize social interaction in other chicks. However, chicks reared in isolation will need to be tested similarly before a definite conclusion can be reached about the role of social rearing.

The experience of social rearing with siblings also channels the imprinting preference, in the natural setting to the hen of the species. In other words, social rearing enhances the choice for the familiar stimulus to which the young have imprinted. This has been shown in ducklings, but not yet in chicks. Johnston and Gottlieb (1985) found that Peking ducklings exposed to a stuffed natural model of a maternal duck followed by being reared with their siblings developed a specific preference for the familiar duck over an unfamiliar redhead duck, whereas ducklings similarly exposed but raised in isolation did not have a preference for either of these two stimuli. Thus, the context of social rearing enhances the visual preference for the familiar duck.

Various degrees of social isolation reduce the preference for the familiar duck. For example, the imprinting preference is strongest in ducklings raised in groups of eight to 12 individuals, and absent in ducklings reared in physical isolation but allowed to see and hear their siblings or reared with only one sibling (Lickliter and Gottlieb, 1985).

The social experience which enhances the imprinting preference must occur after the initial exposure to the model of the maternal duck (Lickliter and Gottlieb, 1987). Social rearing prior to imprinting does not detract from the enhancing effect of social rearing after imprinting, but on its own it has no contributing effect. Moreover, Lickliter and Gottlieb (1988) found that the mallard ducklings need social experience with siblings of their own species if they are to show the enhanced preference. To be raised with a group of domestic chicks was ineffective for enhancing the preference of an imprinted duckling for its conspecific mallard hen. It is, however, possible that the experience of being raised with domestic chicks may enhance an imprinting preference of a duckling previously exposed to a model of a domestic hen. This has not yet been tested.

It is not known how the specific experience of being reared with siblings of the duckling's own species brings about the enhancement of the preference for the familiar, conspecific female, particularly since the maternal visual pattern in this species differs considerably from the visual features of the young, and any broad visual similarities between the adult duck and the young might well be equally common to conspecific ducklings and domestic chicks.

To further complicate matters, young precocious birds also imprint on their siblings or brood mates, and this influences the preference for the adult female when the siblings are in her vicinity. Thus, if socially reared ducklings are given a choice of their familiar duck and a group of siblings, they choose the latter (Lickliter and Gottlieb, 1986). This result raises some interesting questions about imprinting in the natural environment, since siblings are always present with the adult female. When the group of siblings moves off to follow her is it, therefore, always a matter of probability and relative preference? Of course, peer imprinting along with filial imprinting will hold the entire group together, but should peer imprinting be stronger than filial imprinting on the adult female the group might never manage to follow her. It is possible that the balance of these relative preferences changes with the age of the young and with learning experiences. In the experiments of Lickliter and Gottlieb (1986) the ducklings were tested for imprinting preferences at 48 and 72 h after hatching. Socially reared ducklings showed a preference for the duck over the siblings at both of these times, whereas ducklings raised in isolation showed a preference for the familiar duck at 48 h but not at 72 h. It would be interesting to know what happens in older ducklings, and in other precocious species.

For peer imprinting of ducklings, rearing of the ducklings so that they can see and hear other ducklings but not interact directly with them is sufficient to interfere with the imprinted preference for the familiar duck, although not sufficient to redirect the preference to choice of siblings over the duck (Lickliter and Gottlieb, 1986). Peer imprinting occurs if the duckling is reared in contact with a group of stuffed ducklings (Dyer *et al.*, 1989). Physical contact therefore seems to be the most important factor determining peer imprinting.

The shift in preference towards siblings occurs irrespective of whether or not the ducklings receive continuous exposure to the model of the familiar duck, suggesting that other forms of social interaction between the young and a living adult female must maintain following of the adult female by natural broods. Indeed, this also appears to be the case for development and maintenance of species-specific imprinting on the adult female. Dyer *et al.* (1989) raised ducklings with siblings and a model of a mallard hen with which they could interact but only passively (i.e. they could not follow it) and found that in the test at 72 h there was no preference for the familiar duck over a model of a different species of duck. Collectively these studies demonstrate that imprinting in the natural context with the adult female and siblings present is a much more complex process than laboratory studies of imprinting have suggested (Lickliter *et al.*, 1993). The entire process involves imprinting to more than one stimulus and is a prolonged, interactive process of social exchange. This has certainly been shown to be the case in ducklings. We now await a similar series of investigations on domestic chicks. Similar results have, in fact, been obtained with bobwhite quail (McBride and Lickliter, 1993). Chicks of this species develop a species-specific preference for a stuffed model of a bobwhite hen presented together with the maternal call of the species only if they have been socially reared with members of their own species.

It is interesting to consider the effects of social rearing on the imprinted (learned) preference along with the emergent predisposition for approaching the hen. A predisposition represents a growing attraction for stimuli which possess some, or all, of the features of an adult conspecific or, in its absence, for another complex stimulus (Horn, 1985). The predisposition develops without the need for prior exposure to the imprinting stimulus but is unmasked by certain rather non-specific visual experiences. It does not involve prior learning about the complex stimulus *per se*. By contrast, the enhancement of the preference for the imprinted stimulus requires prior exposure to that stimulus and thus prior learning. Although the imprinted preference is enhanced by certain forms of social experience, those experiences are also rather non-specific to the imprinting itself. Thus, the environmental context seems to be designed both to unmask a genetic predisposition for attraction to the hen and to channel learning so that the preference for the hen becomes stronger. The same is true for the development of the preference for siblings. In other words, development under natural conditions with the hen and siblings occurs in an environment well designed to enhance imprinting on both the hen and siblings, or peers. That is, the species shapes its own particular environment in a species-specific manner to enhance the early learning processes essential to survival. As Lickliter *et al.* (1993) have so clearly stated, there are developmental systems, which include both genetic influences and the influence of species-specific features of the environment, that are transmitted from generation to generation. Alternatively, as Kovach and Wilson (1993) have recently demonstrated using quail chicks, the process of imprinting clearly involves genotype–environment interaction and represents genetic canalization of behavioural development.

Lateralization of imprinting and social recognition processes

Domestic chicks show left–right eye asymmetry in making the choice of a companion over a stranger (Vallortigara and Andrew, 1991). When a chick is tested monocularly using its left eye it chooses its companion, whereas when using its right eye it chooses at random. This result suggests that the recognition process is carried out in the right hemisphere, to which the left eye projects most of its information. Apparently, the neural systems, connected to the right eye (mainly in the left hemisphere) attend to chicks as a category, whereas those connected to the left eye attend to small differences which designate individuals (Table 3.1).

Further support for this hypothesis comes from other experiments by Vallortigara and Andrew (1991). They raised chicks in isolation cages with a table-tennis ball, painted red with a white bar on one side, suspended at the chick's eye level. The chicks imprint on such objects. On day 3 they were tested in the runway with the familiar ball suspended at one end and an unfamiliar one, with various rotated positions of the white bar, at the other

Table 3.1 Summary of the functions carried out in a lateralized manner by the left and right hemispheres of the young chick. In most cases these lateralities have been deduced from the eye preferentially used to view a stimulus or the eye systems that are involved in superior performance of each given task, but in some cases there is direct confirmation of the hemispheres involved by the use of intracranial drug treatment. The data presented give an overall picture, although changes in preferential eye use do occur at certain ages (Chapter 5). For discussion of the left hemisphere's involvement in colour discrimination see Chapter 4.

Left hemisphere	Right hemisphere
Categorizes stimuli	Analyses detailed and specific properties of a stimulus
Categorizes conspecifics	Recognizes individual conspecifics
Categorizes food versus non-food	
Responds to large (category-type) changes in the imprinting stimulus	Recognizes small changes in the imprinting stimulus
Stores memory of visual discrimination tasks	
Facilitates changing the visual search image for food	
Controls the right foot, which initiates bouts of ground-scratching	
Encodes colour as an aspect of particular memories (passive avoidance, visual discrimination for food)	
	Attends to topographical cues
	Responds to novelty
	Used to scan overhead in response to an alarm call
Preferred for viewing an imprinting stimulus (hen) and other attractive stimuli	Preferred for viewing less interesting, novel stimuli
Suppresses copulation and attack	Facilitates copulation and attack
	Facilitates fear responses
Attends to auditory stimuli	Attends to olfactory stimuli
Monitors the environment during sleeping	

end. Male chicks raised with a ball with a vertical white bar and tested monocularly using the left eye chose the familiar object when the unfamiliar object had a horizontal bar, but they chose the unfamiliar object when it had an oblique bar at 45° to the vertical. Those tested using the right eye chose randomly between such objects. Only when given a choice between the familiar object and a red ball without a white bar did they choose the familiar object. To summarize, male chicks using the left eye respond to small changes in the imprinting stimulus (which in the natural environment might be the hen or a sibling), and those using the right eye respond only to large changes, which would indicate a category but not an individual. Both forms of processing must be essential for deciding whether to approach or avoid a stimulus. Therefore, in most cases the differential processing by each hemisphere is likely to interact. However, males tested using both eyes behaved like left-eyed males, suggesting that the right hemisphere may dominate for this task, and at this age (Chapter 5). Female chicks chose the familiar model under both monocular conditions as well as when using both eyes, but, when tested with a familiar and an unfamiliar chick, the females using the right eye did not make a choice (Vallortigara and Andrew, 1991). This means that females have lateralization, but they do not always express it.

Left eye and right hemisphere specialization for recognition of familiar conspecifics is also manifested in monocular tests of social pecking behaviour (Vallortigara, 1992b). Social discrimination was measured in terms of pecks directed to brood mates versus those directed to strange chicks. Chicks using the left eye directed more of their pecks to the strangers, whereas those using the right eye did not choose between their companions and strangers, although they pecked just as much. The neural circuits fed by the right eye (primarily in the left hemisphere) respond to chicks as a category, but not as individuals.

As will be shown in Chapter 4, the left and right hemispheres of the forebrain are used differently in encoding an imprinting memory (Table 3.1). The left and right eyes are also used differentially to view a familiar, imprinting stimulus, such as a hen. Chicks usually view large stimuli using the lateral monocular field of vision. Andrew and Dharmaretnam (1991) have devised an excellent test for scoring eye use in such situations. The chick is allowed to view stimuli by putting its head through a small hole and the head angle is assessed from videotaped recordings. Precise fixation angles can be determined. Using this technique they have assessed eye use to view a live hen, a live rat and a small light. On day 8 male chicks view the hen with the right eye and on day 11 with the left eye (Andrew and Dharmaretnam, 1991; Dharmaretnam and Andrew, 1994). This changing pattern of eye use with age presumably reflects a change in the form of analysis of the imprinting stimulus viewed by the chosen eye, or of the environment viewed by the other eye while the imprinting stimulus is monitored by the first eye (Andrew, 1991a; see Chapter 5). Overall, however, males and females preferentially view an imprinting stimulus (a hen) with the right eye. A similar eye preference occurs for viewing a rat, which is clearly a pertinent stimulus and may well be considered a predator by the chick. Workman and Andrew

(1989) have also reported an emergent right eye preference in 8-day-old chicks to view a human in their presence (Chapter 5). Viewing a small light, a novel but not very attractive stimulus, by contrast is preferentially performed by the left eye (right hemisphere).

The overall use of the right eye to view the hen suggests, according to Dharmaretnam and Andrew (1994), a special involvement of the left hemisphere in the analysis of stimuli that may evoke approach and imprinting in chicks. Similar neural systems might also be used to assess predators and the need to withdraw or freeze. If so, the left hemisphere would be involved in decisions of both approach and withdrawal in response to conspicuous, attractive stimuli. This would be a logical organization of brain mechanisms. However, there are some apparent contradictions with other evidence for lateralization. As we will see later, the left eye and right hemisphere play a greater role in fear responses, which one might associate with withdrawal responses.

Olfactory imprinting is also lateralized. By blocking either the left or right nostril of the chick with a malleable wax plug before testing, Vallortigara and Andrew (1994) found that the chicks were able to choose the familiar odour if they used the right nostril (with direct olfactory input to the right hemisphere), but use of the left nostril produced random choice. Thus, the right hemisphere appears to have a dominant role in olfactory imprinting. Note that the left and right nasal passages are completely separated from each other by a layer of cartilage (see Fig. 1.8A), and so blocking one nostril restricts stimulation to the olfactory bulb on the same side as the open nostril.

Sexual Imprinting

By the process of imprinting young chicks learn to recognize their close kin. Filial imprinting allows the chick to recognize its hen and siblings and sexual imprinting determines the choice of a sexual partner in later life (Bateson, 1990). Thus far we have discussed filial imprinting in some detail. Sexual imprinting, or the formation of sexual preferences, has been studied far less often and the more systematic studies have used Japanese quail (Bateson, 1980; 1982) or zebra finches (Ten Cate, 1989b; Kruijt and Meeuwissen, 1991) and not domestic chicks. These studies have shown that individuals prefer a sexual partner to be slightly different, but not too different, from their siblings, thus ensuring optimal outbreeding and thereby avoiding extreme inbreeding or extreme outbreeding (Bateson, 1978, 1983b).

A similar form of sexual imprinting appears to occur in domestic chicks. Females reared with a single male chick, when tested at 3 months of age, chose to spend more time with an unfamiliar male of the rearing strain rather than with the familiar male or an unfamiliar male from a novel strain (Bolhuis, 1991). Presumably, this means that the unfamiliar male of the rearing strain has been chosen as a sexual partner, but this was not scored directly. Only the age of the chicks at testing indicates that sexual, rather than filial, imprinting has occurred.

Sexual imprinting differs from filial imprinting in three main ways. First, it is expressed as sexual behaviour directed to the preferred individual or object, rather than approach or following behaviour. Second, it differs in terms of the age at which it is expressed, and third, in the period of exposure to the stimulus required for the sexual preference to develop. Sexual imprinting preference is expressed in older birds, and it forms during exposure at a later age. Vidal (1980) exposed male domestic chicks to an artificial model designed to allow the chicks to nestle inside it and to be of a shape which would facilitate copulation by the adult. They were raised with this model for 15 days commencing on day 1, 16 or 30. At 5 months of age the now mature cocks were given a choice between this model and a novel model or a conspecific, and sexual behaviour was scored. The birds exposed from day 30 on were found to have developed a strong sexual preference for the familiar model; even though during training they showed only weak filial (following) attachment to it. The other two groups, exposed to the model at an earlier age, showed stronger filial attachment and no sexual preference for the familiar model. Thus, filial and sexual imprinting preferences are clearly separated with respect to the age at which they form. Further study will be needed before we can understand the associations between filial and sexual imprinting, and whether or not they each utilize different neural circuits.

Fear Behaviour

Imprinting and developing fear of novel objects

After the chick has imprinted it no longer approaches a wide range of conspicuous visual objects, but rather shows fear responses to novel stimuli. The increased fear of novel stimuli may develop either as a consequence of imprinting (Hinde, 1955; Sluckin and Salzen, 1961; Bateson, 1964b) or independently as a result of maturation of the escape tendency (Hess, 1959a,b; Weidmann, 1958; also see Chapter 5 for discussion of changes in fear with age and the relationship to the development of brain lateralization, Andrew and Brennan, 1984). Salzen (1962) prefers to see imprinting and the development of fear of novel objects as inseparable aspects of the same neural mechanism. Whatever the explanation for the increased fear of novel stimuli, the occurrence of imprinting serves to keep the chicks in contact with the hen and to ensure that they avoid potentially dangerous novel stimuli or situations.

Fear has also been shown to develop as a consequence of experiences that are particularly related to social rearing or isolation (Kruijt, 1964; Jones,

1987b). As Jones (1989) states, there is no consensus of opinion regarding the age at which fear behaviour first appears in chicks. Immediately after hatching they will avoid frightening stimuli or, for example, crouch when they are presented with a looming visual stimulus. Fear behaviour appears to develop earlier in socially reared chicks, as opposed to isolated ones, and this could be a consequence of imprinting on peers. On the other hand, fear in isolated chicks can be reduced by raising them with a variety of visual and auditory stimuli, or by handling them regularly (Jones and Faure, 1981a; Jones, 1987a; Jones and Waddington, 1992, 1993). Increased visual complexity, for example, reduces fear responses (Jones, 1982). In fact, fear responses to humans by chicks can even be reduced by allowing them to observe other chicks being handled by a human (Jones, 1993). However, it is not yet certain whether this represents a genuine case of observational learning by the chicks, because simply allowing them to see a human standing in front of and touching the cage also reduced their fear of humans.

Jones (1987a) has developed a battery of tests to measure fear in chickens. These include assessing the chick's behaviour in the open field, timing how long it takes to emerge from its home cage into an open field, and timing how long it remains in a state of tonic immobility. Tonic immobility is an unlearned response characterized by the adoption of a catatonic state when the chick is physically immobilized by placing it on its back or side (Gallup, 1974). The more fearful the chick, the longer it will remain immobile in this state. Handling and other experiences of environmental enrichment during early life reduce the duration of tonic immobility and other fear responses. Some researchers have physiologically assessed stress in fear-inducing conditions by measuring the levels of hormones released from the adrenal cortex into the plasma (Beuving *et al.*, 1989). The level of corticosterone, for example, seems to be higher in chicks that are more fearful: there is a trend for higher levels of corticosterone to be present with more behavioural measures of fear (Jones and Merry, 1988).

Fear behaviour changes throughout development in the chick (Chapter 5) and it also varies between different strains of chick (Phillips and Siegel, 1966; Jones, 1977). These issues have much importance in the husbandry of chickens for commercial practice and are intimately related to their welfare. For example, chicks that perform more feather pecking of cage mates are found to be more fearful in the tonic immobility test (Vestergaard *et al.*, 1993). The welfare aspects of fear behaviour and its control will be discussed in more detail in Chapter 7.

There are also sex differences in the responses of chicks measured in many of the tests used to assess fear behaviour. In the open field, males have been scored as less active and showing fewer vocalizations (Jones, 1977). Also, males have a longer latency for emerging from a dark enclosure (Jones, 1979). Although there have been contradictory results for these sex differences (Vallortigara and Zanforlin, 1988), in general it must be said that males are more fearful than females, at least at low to moderate levels of fear (Jones and Faure, 1981b, 1982a). However, Vallortigara and Zanforlin (1988) have presented evidence that the sex differences in the open field result from a greater need of the females to reinstate social contact. They argue that, because there is no sex difference in chicks which have previously been adjusted to social isolation, the shorter latency of and more ambulation by females is not due to fear *per se*, but rather to their attempts to seek social companionship. In fact, female chicks will run faster than males down a straight runway when social reinforcement (a cage mate) is used, whereas males run faster than females when non-social reinforcement (food) is used (Vallortigara *et al.*, 1990).

Whatever the reason for the sex differences in open-field behaviour, it is clear that previous social/isolation experience influences open-field behaviour. Other indicators of fear behaviour might also depend on previous experiences of different kinds. In ducklings, for example, freezing in response to hearing a maternal alarm call for the first time depends on auditory experience around the time of hatching, since ducklings that have been devocalized and socially isolated exhibit reduced levels of freezing (Miller and Blaich, 1988). The freezing response can be reinstated in these ducklings by either playing them recorded sounds of ducklings or by rearing them socially (Miller *et al.*, 1990). Overall, fear behaviour is dependent on both previous experience and the context of testing.

Alarm Calling

Following the topic of fear behaviour in chicks, it is worth discussing some recent research on the responses of adult chickens to predators. Domestic chickens give different alarm calls in response to detection of aerial and terrestrial predators (Gyger *et al.*, 1987). Predators approaching on the ground elicit high amplitude, pulsating calls, whereas aerial predators elicit calls of longer duration with an initial short pulse followed by a longer scream- or whistle-like element. Both cocks and hens given alarm calls to potential predators on the ground, causing the chicks to take cover, but it is largely the cocks that call in response to aerial predators and they do so more readily when they are in the presence of a male or female conspecific (Evans and Marler, 1992), irrespective of whether that conspecific is a chick or an adult (Gyger *et al.*, 1987; Karakashian *et al.*, 1988), or even in the presence of a videotaped image of a conspecific (Evans and Marler, 1991). Therefore, alarm calling to aerial predators, at least, is dependent on the social context, as one might have predicted. It is also dependent on the testosterone level of the cocks (Gyger *et al.*, 1988). Alarm calls to ground predators are also dependent on the presence of an audience (personal communication, C.S. Evans, 1994).

Apparently, conspecifics interpret the meaning of the alarm calls.

Playback of the aerial alarm call leads hens to run to take cover more reliably than does playback of the alarm call given to a terrestrial predator (Evans *et al.*, 1993a). Also, aerial alarm calls evoke crouching behaviour, whereas ground alarm calls lead to the adoption of an erect, vigilant posture. Hens played the aerial alarm call also look upwards more often, even when no aerial visual image is present. Therefore, the alarm signals not only encode information about the internal state of the sender, but also about the circumstances prevailing at the time, as shown previously for vervet monkeys (Seyfarth *et al.*, 1980). The meaning of the information conveyed by the call can be interpreted by the bird receiving it so that appropriate fear or protective responses can be adopted.

Although chickens discriminate between aerial and ground predators, they apparently do not specify between different predators within, at least, the aerial category. In fact, the aerial predator alarm is evoked by small passerine birds as well as raptors, and even by insects or falling leaves, although these results were based on a small sample size (Gyger *et al.*, 1987). The speed and size of raptor-shaped images moved overhead determines the amount of alarm calling: large stimuli moving at high speeds are most effective (Evans *et al.*, 1993b). This lack of specificity may reflect use of a simple perceptual mechanism which develops without learning, although this is not known. It must also be remembered that, although the chickens used in the laboratory studies mentioned had had previous experience of living in outdoor aviaries and therefore some exposure to predators, they were not living in a completely natural environment with full opportunity to learn about different aerial predators. That is, although potential predators were seen, the chickens were not directly exposed to attack. Chicks with previous experience in the natural environment will need to be tested before firm conclusions about the ecological significance of their alarm calling can be made. In addition, it would now be most interesting to study the development from hatching to adulthood of alarm call production, as well the interpretation of these calls by conspecifics of various ages.

Finally, it should be mentioned that the hens played the aerial alarm call looked up by tilting the head so that they could scan overhead using the lateral, monocular visual field, and that they showed a significant preference for using the left eye for doing this (Evans *et al.*, 1993a). This result may indicate preferential use of the right hemisphere to monitor novelty and/or control fear responses, as shown in young chicks (Andrew *et al.*, 1982; Phillips and Youngren, 1986). Phillips and Youngren (1986) reported that placement of a kainic acid lesion in the archistriatal region of the right hemisphere of chicks suppresses their distress calling when in an unfamiliar environment, whereas a lesion in the left hemisphere does not. Andrew *et al.* (1982) found increased distress calling (peeps) in response to presentation of a novel, fear-inducing stimulus to chicks using their left eye and right hemisphere, compared to those using the right eye and left hemisphere.

Learning to Feed

From the time of hatching the young chick pecks at small visual stimuli, preferring three-dimensional, spherical objects (Dawkins, 1968). Its pecking is thus directed towards visual stimuli that have the main properties of the food grains on which they will later feed. Small moving stimuli also attract the attention of chicks of all ages, and this relates to feeding on small insects. Indeed, chicks will actively pursue and peck at insects on the ground and in the air.

Newly hatched chicks also peck more readily at objects of particular colours. It has been reported that their colour preferences for pecking are orange and blue (Hess, 1956) or red and yellow (Goodwin and Hess, 1969), and that these differ from their colour preferences for approach and imprinting (Hess, 1959a). Contrary to these original claims, it is now known that the colours that elicit most pecking also elicit approach most readily (Davis and Fischer, 1978; Clifton and Andrew, 1983). These colours are red and blue. In fact, the ease of eliciting pecking at small beads shows a U-shaped

Fig. 3.3. The chicken uses its beak as if it were a hand. This drawing is a modified version of one for the pigeon by H.P. Zeigler, reproduced with permission.

relationship to wavelength with the highest levels of pecking at either end of the visual spectrum (Andrew *et al.*, 1981). The colour preferences are stronger for pecking a small (5 mm) objects than larger ones (13–37 mm) (Clifton and Andrew, 1983). Thus, particular combinations of colour, size and shape elicit the pecking response of young chicks.

Although the newly hatched chick actively pecks at visual stimuli, much of this pecking is with a closed beak and does not lead to mandibulation or ingestion of the object. Other pecks involve mandibulation but are not followed by ingestion. Some objects which do not give a nutritive reward are swallowed and presumably this is important for the early stages of learning to feed, but the chick can survive on the remains of the yolk sac for the first two to three days posthatching. Most of the pecks made immediately after hatching are therefore likely to be of an exploratory nature. Turner (1964) described pecks with the closed beak as investigatory. In fact, birds must use the beak to explore the environment, much as we use our hands (Fig. 3.3). The beak is also used as a 'weapon' to inflict aggressive pecks, but these are infrequent during the first two weeks posthatching, as will be discussed later in this chapter.

The role of the beak and reward systems

When chicks peck, with a closed or open beak, they receive tactile feedback, and for at least the first two days posthatching, when the yolk sac reserves are still the main source of nutrition, tactile inputs may form a reward system (Hogan, 1973). The nutritional reward system may not develop until day 3 posthatching, or, as Hogan (1971a) stated, the hunger system does not develop until day 3. During the first three or four days of life the chick establishes a broad category of object types suitable for pecking and then swallowing (Hogan, 1973). Together with this development, the amount of pecking increases and also the specificity of cues which will elicit pecking increases (Dawkins, 1968; Hale and Green, 1988). Presumably the proportion of pecks with a closed beak made at potential food objects might decrease as the chick ages and learns to recognize food.

Closed-beak or investigatory pecking must, however, remain important throughout life. In their second week of life, at least, chicks find it rewarding to peck at grains which can be mandibulated or which bounce away when they are pecked. Evidence for this comes from tests which involved depriving chicks of food on different substrates for 3 h prior to testing on an appetitive task requiring them to discriminate grains of chick mash scattered at random on a background of small pebbles adhered to a clear plastic floor (Fig. 3.4; Rogers *et al.*, 1974). After deprivation of food while in a cage with no peckable objects on the floor, the chicks actively search for the grain. At first they are unable to discriminate the grain from the pebbles but within some 60 pecks they almost completely avoid the pebbles and choose to peck at grain, even

Fig. 3.4. A chick searching for grains of mash scattered on a background of small pebbles, which have been adhered to a floor of clear plastic. The pebbles cover the same range of sizes and shapes as the mash and are of a variety of colours overlapping with that of the mash. Therefore, the chick apparently uses texture and/or brightness cues to discriminate the grains of mash from the pebbles. The task is referred to a 'the pebble-floor task'.

though they may ingest very few, if any, of the grains (Fig. 3.5). Chicks deprived of food for 3 h but allowed to peck at and mandibulate coarse pieces of sawdust clippings scattered on the floor, by contrast, do not learn to discriminate grain from pebbles in the 60 pecks allotted (Fig. 3.5; Reymond and Rogers, 1981a), suggesting that pecking preferences are not solely based on nutritive reward. In other words, deprivation of food involves deprivation of non-nutritive as well as nutritive input. This conclusion was further supported by the fact that chicks deprived of food on a clean floor but force-fed using a tube inserted into the crop also learn to avoid pecking pebbles, preferring to peck at the grains even though they were not hungry (Fig. 3.5).

With an extended deprivation period of 15 h, nutritive reward becomes a more important factor controlling pecking, even if the chick is allowed to peck at sawdust during the deprivation period. After this extended deprivation period, all chicks learn to avoid pecking at pebbles (Fig. 3.5).

When a chick pecks with a closed beak it is the upper mandible which contacts the object being pecked because it overlaps the lower mandible, but tactile sensations would be transmitted to both mandibles. Both mandibles are richly endowed with innervation for taste and proprioception (Kuenzel,

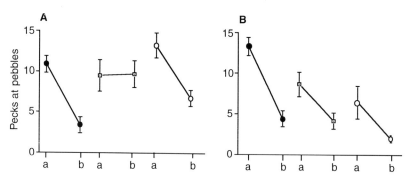

Fig. 3.5. Performance on the pebble-floor task, as in Fig. 3.4, is indicated by plotting the mean number of pecks at pebbles (together with the standard error) in the first and last 20 pecks of the test ('a' and 'b' respectively). The chicks are allowed a total of 60 pecks, each scored as a new choice and not repeated pecks at the same grain or pebble. The data are from three groups of eight chicks tested on day 9 or 10 posthatching after either 3 h (A) or 15 h (B) of deprivation; ●, controls deprived on a clean floor without any peckable objects; □, deprived on a floor covered in coarse sawdust; ○, placed on clean floor for an equivalent time but force-fed so that the chicks were not hungry at the time of testing. Note that all groups except the one deprived for 3 h on a floor of sawdust shift to avoid pecking at pebbles in the last 20 pecks. These results indicate that the chicks prefer to peck at grain for some of its non-nutritive properties.

1989), and furthermore the end of the beak contains a number of cutaneous nociceptors, which transmit the sensation of pain (Breward, 1984, 1986). When the beak is trimmed, a practice involving removal of the tip of the upper mandible rather commonly used in the poultry industry (Gentle *et al.*, 1982), these receptors are removed. Consequently, the rewards associated with pecking are changed. The beak-trimmed chicks peck at the same rate as normal chicks but they shift their preferences towards larger seeds, such as sorghum (Adret-Hausberger and Cumming, 1985). They also swallow a lower proportion of the seeds at which they peck (Workman and Rogers, 1990). The re-shaped beak appears to be inappropriate for food mandibulation and food intake is reduced (Gentle *et at.*, 1982; Blokhuis *et al.*, 1987). Also, the re-shaped beak may be less able to provide tactile, proprioceptive and even taste rewards. These considerations raise both ethical and practical concerns about the practice of beak trimming (Workman and Rogers, 1990).

Beak trimming of adult hens causes at least a temporary decrease in food intake, the body weight being reduced for some six weeks, although pecking rate at first increases and then returns to the preoperative level (Gentle *et al.*, 1982). Feeding efficiency is reduced by beak trimming because the birds either fail to grasp the food pellets or do not transfer them to the pharynx for subsequent swallowing, but the stereotyped pattern of pecking is not altered by beak trimming. It is as if the operation causes something akin to the phantom limb effect which occurs in humans following the loss of a limb.

Age changes in feeding behaviour

Pecking at grain in preference to inedible objects, such as the pebbles adhered to the floor (Fig. 3.4), occurs more readily in chicks in their second week of life compared to the first week. Measurements made on chicks reared in isolation have shown a marked increase in activity and a significant increase in pecking at the ground on day 5 posthatching (Broom, 1969). Apparently, from this age on chicks attend to peckable objects on the ground and must begin to feed for survival as the yolk sac reserves disappear. In group-reared chicks, these behavioural changes occur earlier (Broom, 1969), although one may not expect a difference in the time of yolk sac depletion. Thus, it would seem that social facilitation might speed the onset of each stage of behavioural development.

Social facilitation of feeding

Social facilitation is clearly a major factor in determining early pecking responses (Tolman, 1964, 1968). The amount of food ingested is increased by the presence of a companion chick (Tolman and Wilson, 1965). If the companion is actively feeding, it has a maximal effect of stimulating pecking in the other chick, but by merely being present and active the companion chick also has some effect (Tolman, 1968). Even a rather crude model of a companion which performs pecking-like movements will facilitate pecking (Tolman, 1964). Yet, a hen-like model moving with pecking-like movements is a most effective model for stimulating pecking by the chick (Tolman, 1967a), and the tapping sounds produced by the pecking movement also induce pecking by the chick (Tolman, 1967b). Even without the visual presence of a model, sharp tapping sounds will facilitate pecking (Suboski, 1987). In the natural setting young chicks peck wherever the hen pecks (Collias, 1952). This may be a process by which the hen stimulates the young chicks to feed and teaches them to prefer pecking at certain types of potential food objects.

Stimulus preferences can be formed by observing the objects at which a model is made to peck. Turner (1964) allowed chicks to view, through a glass screen, a model making pecking movements at one of a number of small targets and found that the chicks developed a preference for this particular type of target. This sort of learning results in rather stable food preferences if training occurs in the first three days posthatching, even though the targets are not edible (Suboski and Bartashunas, 1984). Clearly, the chicks learn about food types and develop preferences by watching the pecking of the hen.

Vocalization by the feeding hen attracts the chicks to the site where she is feeding (McBride *et al.*, 1969; Evans, 1975; Sherry, 1977). The hen emits a rapid staccato call, pecks at the food, picks it up and drops it repeatedly. The latter behaviour is known as tidbitting. Sherry (1977) demonstrated that it is the visual, and not auditory, presence of the chicks that leads to this display by the hen. Adult cockerels presented with food also give food calls in the presence of a hen, and the calling rate varies with the quality of the food (Marler *et al.*, 1986) and the motivational state of the cockerel (Evans and Marler, 1994). Hence, the cockerel communicates the presence and quality of the food as well as its state of motivation to feed. The cockerel's food calling is given in the presence of either a familiar hen or an unfamiliar hen, but not in the presence of another male (Marler *et al.*, 1986).

There is specificity in the maternal call to which the chicks respond. Cowan and Evans (1974) imprinted chicks to the maternal call of a particular hen together with a swinging pendulum and then scored their pecking at food in the presence of the pendulum plus the familiar call versus the pendulum plus the same call given by an unfamiliar hen. Pecking was suppressed by the unfamiliar call. In the presence of the familiar call the chicks pecked freely and emitted pleasure calls (twitters). This discrimination between parental calls occurred only when the auditory and visual stimuli were combined; as for auditory imprinting (see p. 82), previous exposure to the call alone was insufficient.

Cowan and Evans (1974) also noted another effect of the maternal call. They found that the position of the source of the sound altered the chicks' responses. When the speaker was located vertically over the chicks' heads, the chicks fed and twittered. When the speaker was positioned horizontally against one wall of the cage, they approached it instead. Thus, in a natural context, calls emitted by the hen at a distance from the chicks should lead them to approach her, as indeed occurs, and, once they are next to her and her calls are now emitted over their heads, they should be stimulated to remain in her vicinity and feed. Their state of satisfaction in the latter context is then indicated by their twitter calls (Andrew, 1964). In turn, these twitters made while feeding evoke more pecking to feed by the chicks (Tolman, 1967a), and thus feeding behaviour by the hen–chicken group is further enhanced.

Recently, McQuoid and Galef (1993) have demonstrated that adolescent Burmese red jungle fowl chicks respond to videotaped images of conspecifics feeding from visually distinctive food dishes by displaying an enhanced preference for food in dishes of the type used by the chicks in the videotape. Both auditory and visual cues provided by the videotape of the chicks feeding played a role in producing the social enhancement of feeding preferences. Also, the most effective stimulus was a videotape of an individual actually feeding from the dish, rather than one either active or immobile near the dish but not feeding. Because the enhancement of preferences persisted over many hours, the authors concluded that they did not merely represent social facilitation, which exerts only short-term effects, but some other longer-term process such as stimulus enhancement. It is an effect which apparently involves memory formation, as shown by Rogers *et al.* (1977) in a similar paradigm using live chicks as teachers and treatment of the observers with drugs that block memory formation.

Thus, there are both immediate and longer-term effects of observing pecking by other chicks, and both the act of pecking itself and the choice of stimulus at which to peck is influenced by the observation. Observation of either a pecking hen or pecking chicks has these effects, but so far no-one has conducted tests to see whether or not the pecking hen is a more effective stimulus than the pecking chicks. It might be suggested that the hen should be a more effective stimulus if she is to function as an efficient teacher of feeding preferences.

Apparently, the state of hunger plays an important role in determining whether social learning will occur by observation of the pecking of another individual. In adult hens, at least, it has been shown that one hen learns to direct its pecks to the appropriate keys which will deliver a food reward by observation of a teacher hen, and that this learning is better in observers that have not been deprived of food compared to those that are hungry (Nicol and Pope, 1993). A satiated observer appears to pay more attention to the teacher. If the same holds true for young chicks, we might predict that learning to feed by observation of the hen would proceed best when the chicks still have available sufficient nutrients from the yolk sac, and are pecking, at least in part, for curiosity rather than exclusively to feed.

Visual mechanisms in feeding

While attending to the food at which the hen is pecking, the chick may establish a searching image for pecking objects sharing given characteristics. As in many other species, chicks readily form a searching image which leads them to pecking in runs at stimuli of a similar type. This is an important aspect of feeding behaviour and essential for breaking camouflage to detect food objects against a background which makes them difficult to see (cf. Croze, 1970, for searching images in crows). The existence of a searching image can easily be demonstrated by giving a chick a choice of grains of two colours both of which the chick is prepared to accept as food (say, red and yellow). The chick will peck in runs of red and yellow rather than at random (Andrew and Rogers, 1972; Rogers, 1974); that is, once it has pecked, say, a red grain there is a greater probability than chance of pecking at another red grain. During a run on any one colour, chicks apparently are searching with a set of central specifications for that colour. A stimulus with cues which match those central specifications is pecked in preference to any other. A switch in attention to another colour requires changing the central specifications (i.e. changing the search image). The rate of change of the central specifications varies with the state of hunger of the chick. A chick that is not hungry pecks with longer runs at inedible objects such as pebbles or less preferred food, whereas following food deprivation the same chick will have longer runs of pecking on its preferred food type (Rogers, 1971).

In these examples we have discussed searching images for grains of different colours. Although colour is a most important criterion of food objects for chicks (Dawkins, 1969), decisions to peck are also based on size and shape cues and searching images may be established for either or all of these characteristics. Although newly hatched chicks display specific preferences for the visual stimuli at which they peck, learning in early life shapes or changes these criteria and establishes long-lasting specifications for food versus non-food.

Both of the visual pathways projecting to the forebrain are likely to be used in feeding (Kuenzel, 1989). Pattern discrimination, colour vision, visual acuity and size discrimination are all used to recognize food objects, and these are all known functions of the tectofugal visual system (Chapter 1). Little is known of the role played by the thalamofugal visual system in feeding, although it almost certainly operates in conjunction with the tectofugal system and may be used to direct the chick's attention to food objects detected in the lateral, monocular field of vision. At least, that function has been suggested by Güntürkün *et al.* (1989) for the thalamofugal visual system of the pigeon, since it receives input primarily from the region of the retina that, in turn, receives input from stimuli in the lateral, monocular field. Güntürkün *et al.* (1989) have suggested that detection of objects in the lateral field by the thalamofugal system leads to the pigeon moving its head so that the object becomes positioned in the frontal field for further viewing using the tectofugal system. Possibly, the same mechanism is used in chickens.

The isthmo-optic visual system (Chapter 1) also plays an essential role in pecking behaviour. It focuses the chick's attention on moving stimuli (Miles, 1972). Efferent signals from the isthmo-optic nuclei to the retinae act to increase the sensitivity of retinal ganglion cells to small objects moving erratically, with a jittery motion. Such movement is characteristic of many insects, and so the feedback loop via the isthmo-optic nucleus operates to assist rapid detection of living food. As already discussed in some detail in Chapter 1, the isthmo-optic pathway is also involved in increasing retinal sensitivity when the chick turns to look into darker areas (Rogers and Miles, 1972). It makes objects in darker areas more visible to the feeding chick. Hence, an insect scurrying into a dark region for safety might be clearly visible to the chick with its well developed isthmo-optic feedback pathway. As the isthmo-optic nucleus is better developed in ground-feeding birds (Uchiyama, 1989), it is conceivable that its function in feeding is primary. However, in addition to being used for detection of prey, it may also be used for detecting predators particularly while the chick is feeding (Chapter 1).

Predator detection while feeding is also enhanced by the structure of the eye. The chick has a ramped cornea and other optical aspects of eye structure that, as in the pigeon (Nye, 1973; Schaeffel *et al.*, 1986), may allow stimuli to be in focus at a short distance in the frontal field, for pecking at the ground, and at the same time at a long distance in the monocular, lateral field; that

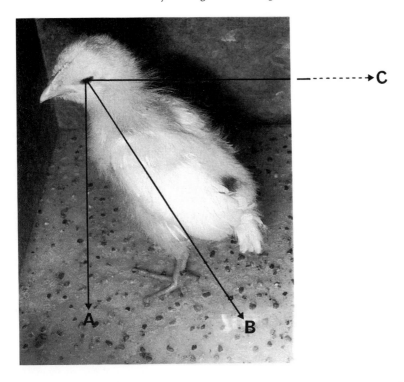

Fig. 3.6. This figure illustrates the way in which the frontal and lower visual field of the chick is focused for near vision (A) with progressively more myopia on moving into the lower field so that images at both A and B would be simultaneously in sharp focus (Martin, 1993). At the same time that the lateral, monocular field is focused for distant vision (C) possibly to detect approaching predators. At B the focus would also be at ground level possibly to detect grains in the next area of search. Adapted from Hodos and Erichsen (1990).

is, in the frontal, binocular field the eye may be myopic relative to the emmetropic lateral field. Hence, while the chick is feeding, its peripheral vision may be focused to detect the approach of predators, from the side or behind. In addition, the lower visual field of the chicken when standing (i.e. the field which looks at the ground) is myopic (short-sighted) relative to the upper field (which views the horizon) so that stimuli on the ground are in focus at the same time as more distant stimuli in the upper visual field (Hodos and Erichsen, 1990). Thus, with its laterally placed eyes the chick has an in-focus panoramic vision, adapted for a bird which walks on the ground and allowing sharply focussed vision of the ground at various distances (Fig. 3.6 and Martin, 1993). In fact, the relative myopia of the lower visual field changes appropriately as the chick grows so that the ground continues to remain in focus as the height of the eye above the ground increases (Hodos and Erichsen, 1990). Thus, the chick's vision is always well adapted for detecting food objects, etc., on the ground simultaneously with navigating in its environment or detecting predators or other objects approaching from a distance. In the lateral field chicks have two preferred angles of viewing (34–39° and 61–66° measured from the beak), presumably utilizing two separate specialized areas of the retina (Andrew and Dharmaretnam, 1993). Distant objects which they wish to examine in detail are fixated at either one of these angles.

Although the chick retina lacks a foveal pit, it does have a central avascular area in which there is a greater density of ganglion receptors (Ehrlich, 1981; Morris, 1987), known as the area centralis. The 61–66° angle of fixation compares to that of pigeons when they use their laterally viewing fovea, whereas the 34–39° angle may use that part of the retina specialized for use in binocular viewing when the eyes are converged but at 34–39° from the beak when the eyes are diverged for lateral viewing (Andrew and Dharmaretnam, 1993).

Chicks are able to accommodate (that is, alter the focus of) each eye independently in the lateral field and, in part, also in the frontal field of vision (Schaeffel *et al.*, 1986). Given their laterally placed eyes, this ability permits a certain degree of independent use of the monocular visual fields (and the respective forebrain hemispheres to which they connect). Furthermore, chicks can move their eyes independently to some extent and thus scan the environment independently with each eye (Walls, 1942; Wallman and Pettigrew, 1985). Their view of the world is indeed very different from that of other species with frontally placed eyes, as well as coordinated eye movements and accommodating abilities.

Lateralization of feeding behaviour

Lateralization of hemispheric use is also a feature of feeding behaviour (Table 3.1). If the chick is tested monocularly on the pebble floor task (Fig. 3.7), it learns to categorize the grain as different from the pebbles when using its right eye but not when it is using its left eye; this occurs provided that the embryos have been exposed to light during the last few days of incubation (Chapter 2). The asymmetry in food/non-food categorization is apparent in chicks tested in their first week of life posthatching (Mench and Andrew, 1986), as well as in their second week of life (Zappia and Rogers, 1987). At least in the second week of life the asymmetry measured is greater in males than in females (Zappia and Rogers, 1987).

It is possible that this asymmetry in the ability to categorize food from non-food objects stems from the fact that the right eye has more thalamofugal inputs to the forebrain (Chapters 1 and 2). The right eye sends its input to the left side of the thalamus and from this side of the thalamus there are more contralateral projections to the right hemisphere. Yet, pharmacological studies would indicate that the left hemisphere is more involved in controlling

Fig. 3.7. A chick wearing a patch over its right eye while being tested on the pebble floor task.

performance on the pebble floor. Injection of a drug which disturbs glutama-tergic neurotransmission (e.g. glutamate itself, a glutamate analogue or cycloheximide) into the left hemisphere on day 2 of life posthatching (of chicks which have hatched from eggs exposed to light) subsequently impairs the ability of binocularly tested chicks to discriminate grain from pebbles (Rogers and Anson, 1979; Howard *et al.*, 1980), whereas similar treatment of the right hemisphere has no such effect. The latter chicks learn as well as untreated, or saline-treated controls. It seems, therefore, that the neural circuits involved in performing this task are located in the left hemisphere.

Other visual learning tasks requiring categorization of visual stimuli are similarly performed better by chicks using the right eye. Gaston and Gaston (1984) trained chicks, aged 3–35 days, in an operant visual learning task with cross and triangle patterns, and found better learning in those using the right eye compared to the left. Memory retention of the task was also superior when the chick used the right eye (Gaston and Gaston, 1984). Although there are inputs from each eye to both hemispheres, albeit minor to the ipsilateral hemisphere, these results, together with the effects of unihemisphere glutamate treatment, indicate that the learning takes place in the left hemisphere only. When the chicks were trained binocularly and then tested monocularly for retention, no recall was apparent in those using the left eye, whereas those using the right eye showed full recall. The memory of the visual discrimination task is apparently stored in the left hemisphere only.

Pecking responses for small, insect-like stimuli are also lateralized in chicks, but the laterality varies with the age of the chick. If the chick is presented with insects simultaneously in both monocular visual fields, by impaling a dead insect (a *Tenebrio* beetle) on each end of a two-pronged fork and then moving them into the visual fields from behind the chick's head, the chick pecks preferentially at the insect in the right field if it is tested on day 8 and in the left field if it is tested on day 10–15 (Andrew and Dharmaretnam, 1991; see Chapter 5). Thus, the hemisphere used to direct feeding responses of this type varies with age, and therefore possibly the manner in which the chick analyses the task.

The important role of the left hemisphere in processing the information for visual discrimination or categorization of food targets is, in fact, counterbalanced by a role of the right hemisphere in processing the topographical cues associated with feeding (Table 3.1). Rashid and Andrew (1989) demonstrated this by training chicks to find food buried under sawdust using either proximal or distant visual cues and then testing them monocularly. The training occurred in an open arena with the food buried in a patch marked by a nearby visual stimulus. The distal cues were placed on the side walls of the arena. At test, no food was present and the relative positions of the visual cues were changed. They were separated from each other by rotating the arena. Those chicks tested using the left eye, from day 9 of life on, used an efficient searching strategy. They spent most of their time searching in the two areas specified by either the proximal or distant cues. The chicks using the right eye searched inefficiently over the entire arena. It is thus likely that the neural systems fed by the left eye (primarily in the right hemisphere) attend to and acquire knowledge about the layout of the environment and where food is likely to be found.

Other tests, not involving feeding *per se*, confirm this involvement of the left eye and right hemisphere in dealing with spatial information. For example, if the chick is first habituated to pecking at a novel, coloured bead and then the position of the bead is changed, by introducing it into the cage from a different place and at a different direction, dishabituation occurs when it is viewed by the left visual field, but not when it is viewed by the right field (Andrew, 1983, 1991b). That is, only the left eye systems respond to a change in positional cues.

The same involvement of the left eye (right hemisphere) with spatial cues and the right eye (left hemisphere) for categorization of stimuli has been demonstrated in another testing situation. Vallortigara *et al.* (1988) trained young chicks to locate a food source by using either positional or colour cues. When males were trained using the spatial cues provided by positioning a small box either to the left or right of the entrance into the testing cage, their learning performance was better when the box was placed on their left side. In a similar task which used colour discrimination as the cue for the food source, the chicks learned better when the reinforced colour cue was placed on their right side (Vallortigara, 1989).

Considering that learning to feed not only requires the chick to categorize food and non-food objects but also to locate the sites at which food is present, we can deduce that both eye systems and both hemispheres are used in feeding behaviour. Therefore decisions to peck to feed must depend on cooperative analysis of different information in each hemisphere.

The analysis of olfactory cues used in feeding might also be lateralized to the right hemisphere, if one can extrapolate from the evidence for a greater degree of involvement of the right olfactory epithelium (and right hemisphere) in olfactory imprinting (see p. 8). Such lateralization has not yet been investigated for feeding behaviour.

Lateralized motor behaviour is also present in feeding chicks. In a study of chicks of a feral strain, originating from North West Island, Queensland, Australia, Rogers and Workman (1993) found a right-foot preference for initiating scratching of the ground. Ground scratching is part of feeding behaviour as it serves to uncover hidden grain and insects. It typically employs raking the ground with each foot alternately, commencing with the right foot. Such right 'footedness' is consistent with the right-eye system's specialization for discriminating food from non-food on the basis of the visual cues of the food objects themselves. Thus, food uncovered by the initial foot movement might be detected and pecked at most efficiently to the right side of the chick, although this has not yet been measured directly. Nevertheless, as the right foot is controlled by the left hemisphere and the left hemisphere categorizes food versus non-food, it is clear that functionally related perceptual and motor activities are located on the same side of the forebrain.

Finally, it should be mentioned that the left hemisphere also plays a dominant role in switching the searching image from one set of specifications to another, at least when the chick is using colour cues (Table 3.1). Control chicks were raised so that they developed a preference for yellow grains of mash but would also eat red ones. When tested in the second week of life with a choice of red and yellow grains, they pecked in runs on either colour. The mean run length on red food scored in an experiment testing 12 chicks was $22 \pm$ a standard error of 7. Following cycloheximide treatment of the left hemisphere on day 2, the mean run length increased significantly to 51 ± 10 pecks (Rogers, 1980). No significant change in this value occurred for chicks treated with cycloheximide in the right hemisphere; their mean run length was 29 ± 9. Thus, disrupting the development of the left hemisphere by cycloheximide causes less frequent switching of the search image from red to yellow grains and vice versa. The drug treatment of the left hemisphere may have its effect by shifting dominance to the right hemisphere, which may be more attentionally persistent. Attention switching between potential food types may be an effective form of food searching when the chick is pecking at grains located in a particular area. In such a situation it may be advantageous to switch between grain types in this area, using the left hemisphere. Yet, when spatial processing concerned with locating a food source is required, the right hemisphere may assume control and in this case it may be advantageous to hold attention more persistently until the food source is located. If so, use of the left and right hemispheres during feeding may relate to the particular demands of the task at any one time during food searching. That is, food searching may commence with use of the right hemisphere to focus attention for a persistent search for the locality of the food source, followed by use of the left hemisphere once pecking begins at that site.

Olfaction in feeding

The olfactory system of birds has an extensive projection to the dorsolateral telencephalon, which is recognized as the pyriform cortex (Reiner and Karten, 1985). Despite earlier beliefs that the olfactory system played only a vestigial role in controlling the behaviour of birds, there is now increasing evidence that it is involved in feeding behaviour (Jones, 1987c; Kuenzel, 1989; Avery and Nelms, 1990) and, at least in species other than the chicken, it is known to influence sexual behaviour (Balthazart and Schoffeniels, 1979) and parental behaviour (Cohen, 1981). The latter will be discussed in Chapter 6.

Earlier in this chapter the role of olfaction in imprinting in the domestic chick was discussed. The role of olfaction in feeding behaviour was demonstrated originally by using the drastic, and rather crude, technique of surgically removing the olfactory bulbs of 4-month-old chicks and noting the effect on food intake (Robinzon *et al.*, 1977). Removal of the olfactory bulbs caused a marked increase in food intake. The authors explained the effect on feeding as being secondary to changes in the activity of the thyroid gland. They did not consider that food intake might have been influenced directly by the absence of olfactory cues from the food itself. The odour of food is, in fact, detected by chicks. Jones (1987c) presented young chicks with mash adulterated with a small amount of orange oil and found that initially they avoided eating it, although orange oil is not an adversant. The chicks were showing neophobia to the unfamiliar odour. Recently, Burne and Rogers (1994b) have investigated the influence of different odours on the pecking responses of chicks in the author's laboratory. Chicks tested in a familiar environment and presented with a novel odour show a decrease in pecking rate. This response depends on the particular characteristics of the odour, with some unpleasant odours, such as garlic, producing a greater reduction in pecking rate than others, such as nesting material. In fact, their behaviour indicates that they have sensed the novel odour because they circle the rearing cage and show increased beak wiping and head shaking. It is, of course, possible that the odour has been detected by the endings of the trigeminal nerve rather than the olfactory epithelium.

Chicks clearly respond to odours when they are feeding. Jones and Gentle (1985) were able to show that they establish preferences for areas treated with familiar odorants. It is, in fact, possible to train chicks to respond to certain odours using conditioning techniques. Using such a paradigm, Stattelman *et al.* (1975) determined olfactory thresholds for pentane, hexane and heptane in pigeons, quail and chickens. The chickens were the most sensitive to pentane and hexane, whereas the pigeons were most sensitive to heptane. Thus there are species differences in olfactory sensitivity to particular odours. It would be of interest to use this procedure with those naturally occurring odours which the chick is likely to encounter both in early life and in adulthood.

The responses of chicks to the odour of blood fall into the category of natural behaviours. Jones and Black (1979) scored the responses of 7-day-old chicks to a dish containing blood, a solution of a red or blue dye, or water. Blood from a conspecific induced avoidance and fear behaviour in the chicks, and it did so more when it was presented in an open dish rather than in a sealed one. The solution of red dye elicited less fear and aversion even though it was visually similar to the blood. The results therefore indicated that the olfactory cues of the blood were important for eliciting the avoidance response.

Other odours also act as warning signals to chicks. Pyrazines are used as warning odours by a number of insect species and they are also produced by moulds and other biological agents of decay. Guilford *et al.* (1987) demonstrated that naive, newly hatched chicks would avoid water contaminated with a pyrazine compound. The chicks were also able to form a conditioned aversion to the pyrazine when it was paired with quinine sulphate, and they apparently could detect the fluid containing the aversant from a distance, suggesting that they were using olfactory rather than visual cues. The results support the notion that pyrazines are effective warning signals used by insects against avian predators.

Chicks show neophobia to certain odours, and they form preferences for odours to which they are familiarized. For example, they show a preference for a soiled substrate from their own home cage to that from a strange conspecific (Jones and Faure, 1982b). When tested in a Y-maze they chose to enter the arm containing their own odour. A similar preference develops for an artificial odour (orange oil impregnated in wood litter on the floor) present in the home-cage of group-reared chicks (Jones and Gentle, 1985). The familiar odour proved to be attractive in an otherwise novel and frightening situation. In the presence of the familiar odour, the chicks were less fearful in the open field. As we have mentioned, the familiar odour must also be present for normal feeding behaviour to proceed.

These experiments demonstrate that, contrary to original views, olfaction is important to the chicken, and deserves more attention by researchers in future.

Taste in feeding

Although the sense of taste in the domestic chicken is far less well developed than that of mammals in general, there is evidence that it plays a role in feeding behaviour. Early reports claimed that there were only some 12–24 taste buds on the tongue of the chicken (Kare and Rogers, 1976), but more recently around 300 taste buds have been found within the buccal cavity (Ganchrow and Ganchrow, 1985). It is now known that the taste buds are distributed on the upper beak, lower beak and mandibular region posterior to the tongue, as well as on the tongue. Sectioning the afferent nerves from these taste buds reduces the chicks' ingestion of both familiar and novel food (Gentle, 1971).

Although grains are insoluble and they are not broken or dehusked by the chick before swallowing, they do have some taste. More likely, however, insects and aversive-tasting objects stimulate the taste receptors of the chicks (Lindenmaier and Kare, 1959). Among other things, the newly hatched chick must learn to avoid pecking at faeces and inedible insects. The response to the presence of a distasteful stimulus in the mouth is to show a disgust response including shaking the head and wiping the bill on the floor.

The fact that day-old chicks learn rapidly to avoid pecking distasteful stimuli has been used in the development of a one-shot learning task known as the passive avoidance bead task. The chick is presented with a coloured bead (usually red) coated in methyl anthranilate, which has a bitter taste. It pecks at the bead and then shows the disgust response. When presented again with a bead of the same colour it avoids pecking at it. This task is used extensively in studies investigating the biological correlates of memory formation, as will be discussed in Chapter 4.

The chicken appears to have the ability to taste water, and this plays a role in its water intake (Gentle, 1975). Chickens also show preferences for some sugar solutions and aversion to sodium chloride and acid solutions, depending on their concentration (Gentle, 1975). In general, apart from its relative lack of sensitivity, the gustatory system of the chicken is similar to that of mammalian vertebrates.

Social Hierarchies in Young Chicks

It was generally accepted that chickens begin to develop a social rank order when they are five to six weeks of age (Guhl, 1958; McBride *et al.*, 1970). This is correct provided one bases the social hierarchy solely on the outcomes of agonistic encounters (Dewsbury, 1982, 1990) because very young chicks show little or no aggressive pecking. Rushen (1984) has reported that males of a domestic chick strain first display aggressive pecking in the second week of posthatching life and reach adult levels of this behaviour by eight to nine weeks. In Burmese red jungle fowl Kruijt (1964) detected the first aggressive pecking at 10 days old, followed by the beginning of juvenile fights at three weeks of age. Therefore, if rank order in chicks is measured by scoring aggressive pecking among the individuals in a group, it is not possible to determine a social hierarchy in the first few weeks posthatching. However, rank order may be revealed by competition for resources, as commonly used to determine social rank in other species (e.g. Friend and Polan, 1978). Indeed, it is known that high-ranking, adult fowls gain priority access to food in competition with the rest of the flock and have priority rights for copulation and for a roosting site (Gottier, 1968). Thus, social rank order may be assessed on the basis of competition for food or other resources.

By measuring competition in groups of eight domestic chicks for access to a food dish and the order of passing through a small gate to enter the feeding compartment, it has been possible to demonstrate that social hierarchies exist even in the first week of life (Rogers and Astiningsih, 1991). The chicks were tested daily from day 4 to day 18 and the scores were compared by subdividing this into three time periods (days 4–8, 9–13 and 14–18). There were changes in position in the social rank order over these time periods, but individuals ranking in the first and second positions maintained their positions across tasks and age. Individuals in the middle and lower ranks varied their positions more, but they showed some consistency over tasks and time. Additionally, a third and very different task not requiring food reward was also used to score competition success. This task, called the chase test or 'worm running' test, involved competition by the chicks to pick up and run with a small (3 mm worm-like) piece of paper. The protrusion of a worm-like stimulus into the visual field directed away from the beak elicits running behaviour in the chick (Hogan, 1966), and others in the group follow and compete for the object. This behaviour was first described by Spalding (1873) and later by Kruijt (1964), who suggested that it has adaptive value in ensuring that large pieces of food will be torn up in the tussle to grab and run with it. The social rank order was determined using this running behaviour by scoring the number of times an individual had hold of the object. Rank order determined in this way tended to be similar to that obtained in the two tests for food competition. Hence, although chicks do not form an absolutely rigid social hierarchy in the first three weeks of life, they do maintain an ordered group structure. In fact, there is rather surprising stability in the highest ranking positions for access to a range of resources over time.

The reader will recall that in Chapter 2 we discussed the way in which light exposure of the embryos influences the stability of the social hierarchy determined by competition for access to a food dish. Compared to groups of chicks incubated in darkness, those that have received exposure to light form more stable rank orders.

It is not yet known whether these social hierarchies determined for young chicks are similar to those determined by scoring aggressive pecking in older chicks. Yet, the level of testosterone circulating in the young chick may influence its position in the hierarchy. At least, testosterone treatment of a low-ranking chick causes it to move up the hierarchy to a higher ranking position, based on the same measures of competition for food or 'worm-running' (Rogers and Astiningsih, 1991). The testosterone treatment is known to elevate agonistic behaviour (Andrew, 1966, 1975a), but the tests used to score rank did not include aggressive pecking. Nevertheless, agonistic encounters outside of the testing time might establish a hierarchy that becomes manifest in competitive interactions. In fact, Zayan (1987) has shown that adult chickens learn and remember social patterns developed from agonistic encounters. Alternatively, the testosterone may exert its affect on competition via its well-known ability to cause attentional persistence or focused attention on a goal (Andrew and Rogers, 1972; Archer, 1974; Rogers, 1974; see pp. 173–176). A focused attention on the food source may well lead to more successful competition and a higher ranking score.

Another way of measuring the social hierarchy in chickens is to assess attack leaps. When a chicken performs an attack leap, it jumps off the ground thrusting both feet at its opponent. When two birds encounter each other, leaping occurs prior to pecking. Rajecki (1988) socially isolated adult, male domestic chickens for either eight days or eight weeks and then tested them in pairs. The hierarchy obtained by scoring attack leaps during these encounters reflected that obtained from aggressive pecking. Even very young chicks can be ranked according to a leap order. Rajecki (1988, 1991) determined hierarchies in very young chicks according to the 'leap order', and, whereas early peck orders did not predict later leap orders, early leap orders did predict later leap and peck orders. Thus, leap order appears to be a very useful measure of the social hierarchy across the life-span.

Young chickens have been shown to adopt another form of social structure based on leadership. Adret-Hausberger and Cumming (1987a) observed in chicks, aged from two to three weeks, strong tendencies for some birds to initiate more feeding activities than others. The leading bird tended to be followed by all of the other birds in its group. They also found that an older bird is accepted readily as a leader by younger birds, hatched in an incubator and without previous experience of an older chick. The young chicks follow the older one to feeding sites and imitate its behaviour (Adret-

Hausberger and Cumming, 1987b). Thus, peer attraction may interact with social learning and leadership.

There are gender differences in the responses that chicks give towards their peers and to strangers. In approach–response tests, females show shorter latencies to approach when they are tested with their peers than when tested with strangers, whereas males show the opposite effect (Vallortigara, 1992a). When given a simultaneous choice between a cage mate and a stranger, a female spends more time with the cage mate and a male spends more time with the stranger. Both sexes direct aggressive pecking towards strangers, but males do so more than females. According to Vallortigara (1992a), female chicks form stronger social attachments than do males. Consistent with this idea, female chicks have been shown to move out of sight of the hen less often than male chicks (Workman and Andrew, 1989). These gender differences apparent in early life may carry over into adulthood, as the feral fowls of North West Island, Queensland, Australia, have a social organization in which a single dominant cock patrols a rather large territory in which a closely knit group of females resides (McBride *et al.*, 1969).

Sleep

There has been relatively little study of sleep in birds, but their EEG patterns are known to be similar to those of mammals (Ookawa, 1971; McFarland, 1989). Birds adopt typical sleep postures and during sleep their threshold for arousal is raised (Amlaner and Ball, 1988). Adult fowl perch for sleeping and adopt a typical immobile posture with the eyes closed and head lowered under a wing (Blokhuis, 1984), but young chicks adopt a variety of postures for sleep. They may show total loss of muscle tone and lie prostrate on the ground or they may sit or stand with the head dropping as they sleep. EEG recordings from surface electrodes on the brain indicate states of quiet sleep, with slow waves of high amplitude, and active or paradoxical sleep, in which the waves are fast and of lower amplitude (Peters *et al.*, 1965; Amlaner and Ball, 1988). It is also possible for birds to sleep unihemispherically, with one eye open and the other closed. In such cases, the hemisphere contralateral to the open eye has an awake EEG pattern, whereas that opposite the closed eye has a sleeping EEG pattern (Peters *et al.*, 1965; Ookawa, 1971; Amlaner and Ball, 1988; Ball *et al.*, 1988). Similar unihemispheric sleep has been recorded in a number of aquatic mammals, particularly in dolphins and seals (Oleksenko *et al.*, 1992; Mukhametov, 1987). In aquatic mammals and in birds, unihemispheric sleep may allow migration to continue while half the brain sleeps. It is also possible that unihemispheric sleep in birds allows vigilance for predators because one side of the brain can sleep while the other remains awake. Prior to hatching much of the EEG sleep pattern is either unihemispheric or in only part of a hemisphere (Peters *et al.*, 1960), which suggests that it may be related to the

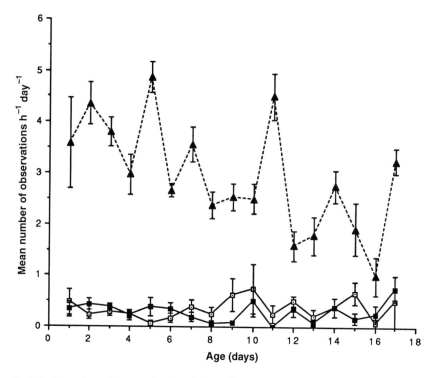

Fig. 3.8. Binocular and monocular sleep (indicated by eye closure) has been scored at 5 min intervals over two 3 h periods during the day on a daily basis for chicks aged up to 17 days posthatching. Means and standard errors for a total of 16 chicks are presented. ▲, binocular sleep; □, left eye closure; ■, right eye closure. Note the decline in binocular sleep with increasing age and the relatively constant, albeit lower, levels of monocular sleep.

neural plastic phases of development although, of course, perceptual stimulation may also play a role in generating the embryo's patterns of sleep and wakefulness.

As mentioned in Chapter 1, the chick has developed sleep–wake cycles of EEG activity by the time it hatches (Corner *et al.*, 1973). When first hatched, the chicks sleep much of the time but with increasing age the amount of time spent asleep declines. Figure 3.8 presents data for day-time sleeping in chicks over the first 17 days posthatching (Rogers and Chaffey, 1994). There is a gradual overall decline in the time spent in binocular sleep (both eyes closed), but noticeable peaks in sleeping are present around day 2 (compared to a significantly lower level on day 4), on day 5, possibly on day 7 and particularly on day 11. The data in Fig. 3.8 indicate that yet another peak may occur on day 17, but older chicks need to be tested to clarify this. As will be discussed in Chapter 5, these peaks coincide with or follow particular transitions in behavioural development. In brief, the peak around day 2, for example, coincides with the optimal time for imprinting, day 5 is when the chicks begin to follow the hen actively and day 11 follows learning about topographical aspects of the environment and an increase in independent activity of the chicks (Chapter 5). Of course, there are other ages at which behavioural transitions occur, but the increased sleeping at these times may perhaps represent maturational requirements of the brain at these particular ages, or alternatively it may occur as a consequence of learning which occurs at these ages. Recently, studies in my laboratory and by Solodkin *et al.* (1985) have shown that filial imprinting leads to an increase in total and paradoxical sleeping by chicks. The function of sleep is far from understood, but studies have shown that in rats sleeping patterns affect learning ability (Ambrosini *et al.*, 1993) and learning leads to an increase in rapid eye movement or paradoxical sleep (Rotenberg, 1992). Indeed, a number of studies using mammals point to a role for rapid eye movement sleep in memory formation (Winson, 1993). Admittedly, these studies do not report that learning causes an increase in total quiet sleep, as measured in chicks, but drugs which affect the receptors involved in neural plasticity and memory formation increase both kinds of sleep (Stone *et al.*, 1992) and age is also a factor in sleeping patterns (Stone and Gold, 1988; Arankowsky-Sandoval *et al.*, 1992). That is,

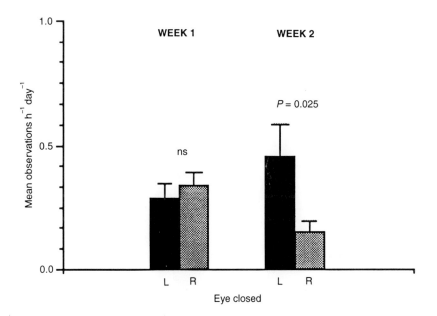

Fig. 3.9. Monocular sleep in the first and second weeks of posthatching life is presented according to the same sampling methods as Fig. 3.8. Note the occurrence of significantly more events of left eye closure compared to right eye closure in the second week. The asymmetry is caused by a significant decrease in sleep with right eye closure from week one to week two (*P* = 0.021).

memory formation may possibly be linked to different forms of sleep in young compared to older animals.

As shown in Fig. 3.8, the amount of monocular sleep in chicks remains relatively constant over the first two weeks of life. At first only about 25% of sleep is monocular but, as a result of the decline in binocular sleep with increasing age, it increases to around 50% towards the end of the second week (measured in small groups of chicks housed in a quiet laboratory).

Moreover, in the second week of posthatching life monocular sleep with left eye closure occurs more frequently than that with right eye closure (Fig. 3.9). This means that overall the right hemisphere sleeps more than the left. A detailed analysis of when right unihemispheric sleeping is more frequent revealed that it occurs from a state of binocular sleep and leads the chick back into binocular sleep or into the awake state (Rogers and Chaffey, 1994). It is transitions from binocular into monocular sleep which occur more commonly by opening the right eye. Transitions from the awake state into monocular sleep occur equally often by closure of the left or right eye (Fig. 3.10). In other

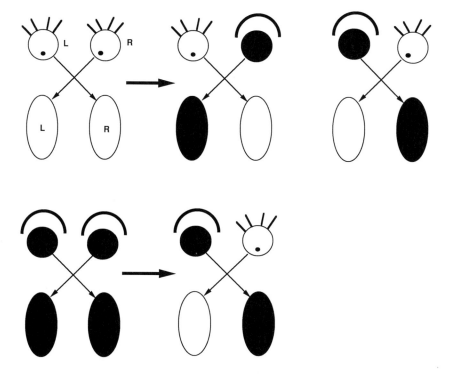

Fig. 3.10. Transitions into monocular or unihemispheric sleep. When the chick enters monocular sleep from the awake state, it is equally likely to close the left or right eye. When it enters monocular sleep from a state of binocular sleep, there is preference to open the right eye and thus attend to the environment with the left hemisphere only.

words, the binocularly sleeping chick has a bias for opening the right eye to monitor the environment. Presumably if nothing has changed the chick goes back to binocular sleep, but if there has been a significant change in the environment it wakes up. It is, in fact, the neural systems fed by the right eye that respond to major stimulus changes, whereas the left eye attends to changes in detail (Vallortigara and Andrew, 1991; see pp. 87–89).

Cycles of Behaviour

The data on sleeping discussed in the previous section were all collected during the daytime using chicks housed under conditions of constant light by an incandescent light bulb positioned above each cage. Even under these conditions chicks show cycles in behaviour. For example, their performance on the pebble floor task varies according to a diurnal cycle (Reymond and Rogers, 1981b). From 1000 h to 1800 h they learn to avoid pecking pebbles within the 60 pecks allowed following 3 h of deprivation of food on a clean floor (see data in Fig. 3.5). Outside this period, however, they peck at grain and pebbles randomly throughout the 60 pecks; that is, at night they show no shift towards pecking for grain in preference to pebbles.

Although these chicks were housed with continuous light, there were other events that occurred with a daily regularity that could serve as external stimuli for entraining their internal clocks (a stimulus that performs this function is said to serve as a *zeitgeber*; Aschoff, 1960). In particular, the time at which the cages were cleaned and fresh food was given served as one of these signals. The diurnal pattern of performance on the pebble floor was characteristic of chicks fed at 0930 hours, but in another group fed at 2000 hours the shift away from pebbles to a preference for grains occurred when they were tested at 2400 hours as well as in the middle of the day (Reymond and Rogers, 1981b). The late time of feeding was followed by a brief period during which they pecked preferentially for food objects. In fact, the chicks fed at 0930 hours had one peak in general activity around the time of feeding and a coincident peak in the amount of food intake. Those fed at 2000 hours had two peaks in general activity, one around the time of feeding and another in the morning at the same time as the other group, and two coincident peaks in food intake. The morning peak could have been synchronized with other signals in the laboratory, since these chicks were not housed in a soundproof room, but other studies have shown that chicks incubated and then housed under controlled, constant conditions retain certain cycles in behaviour, although not exactly according to a 24-hour (circadian) clock. Apparently, there is an endogenous biological clock operating to influence activity and feeding cycles. This, of course, does not mean that it normally operates independently of external cues. The light–dark cycle has been shown to influence feeding patterns in adult fowls, leading to seasonal variations in the time of feeding peaks (Wood-Gush, 1971). However, this behavioural cycle is not completely synchronized to the light–dark signals.

Biological clocks or oscillators with a circadian rhythm may occur in various regions of the central nervous system and so control the various 24-hour cycles in behaviour. In fact, there is evidence for a circadian oscillator in the retina, controlling the function of the rod photoreceptor cells (Schaeffel *et al.*, 1991). The rod photoreceptors are used under conditions of low light intensity, particularly for detecting movement. Their spectral sensitivity extends into the ultraviolet region of the spectrum. In chickens they barely function during the day, but at night they are turned on endogenously with the result of improving night vision. The differing visual abilities, cycling between cone-dominated function in the daytime to rod-dominated function at night must determine at least some of the cycles in visually guided behaviour.

Concluding Remarks

The chick hatches with a well-developed brain, immediately able to make decisions and to form memories. Within the first days after hatching many memories are formed which are crucial for the chick's survival, and each of these has a more or less precisely timed sensitive period during which the learning must occur, although exactly when each sensitive period occurs depends on past experience, even prior to hatching, and other constraints including the stage of development of motor abilities. For example, although it is possible to visually imprint a chick within hours of hatching, the optimal time occurs somewhat later, by which time the chick is able to stand up and approach the imprinting stimulus. In the natural setting therefore, the onset of the sensitive period for filial imprinting in the visual modality is not determined by the ability of the chick's brain to learn about the imprinting stimulus, but rather by the chick's motor abilities and its attention to potential imprinting stimuli. It must be remembered, however, that imprinting in the auditory and olfactory modalities may have occurred well in advance of visual imprinting, even prior to hatching. In fact, learned preferences in these modalities may guide the chick's visual attention to the hen; that is, prior imprinting to her odour and her cluck call may focus the chick's visual attention on the hen. In addition, prior visual experience of an apparently non-specific kind may influence visual imprinting after hatching, as also do social interactions of various kinds.

Most evidence on auditory imprinting indicates that it does not form such a long-lasting memory as does visual imprinting, and so may serve merely to focus attention on the visual imprinting stimulus. Evidence for olfactory imprinting is presently so sparse that no conclusions can be arrived at about its role relative to visual or auditory imprinting.

Bateson (1972) has described visual imprinting as a learning process analogous to painting a portrait in which the broad outlines and salient features are sketched in first to be followed by gradual filling in the details, including the background. In human memory formation, this filling-in process is referred to as adding 'depth' to the memory. In the next chapter we will see how adding depth to the imprinting memory may involve recruitment of extra neural circuits. I would like to expand Bateson's analogy of a portrait to one including sound and odour as well. It may be that the first outline of the 'portrait' is in the form of a 'tone poem' and/or its olfactory equivalent, and that visual imprinting is painted upon this particular outline or framework.

There is a sequence of incorporation of the various sensory modalities into the imprinting memory which apparently reflects their sequence of development, but one should not ignore the complex interactions between modalities. We are only just beginning to understand that it is incorrect to consider each sensory modality as a separate entity. To take up the earlier analogy again, it is as if the 'tone poem' is already placing colours on the pallet or canvas before the painting can begin. These cross-modality interactions deserve much more attention in future research on imprinting.

Other forms of early learning by the chick appear to fall within sensitive periods that may or may not be linked with that of imprinting. Imprinting occurs when the chick can still survive on the remaining nutrients from the yolk sac, and so it does not exactly coincide with any essential need for food ingestion. In fact, the chick's need for food intake follows immediately after the sensitive period for imprinting. This is not to say that learning to discriminate food from non-food has been completely postponed until after imprinting, as it may well have been slowly occurring over the same time period. Nevertheless, it would be worth considering the possibility that visual imprinting and visual discrimination of food involve linked processes separated in developmental time. In fact, some evidence that will be presented in the next chapter indicates that, if visual imprinting is delayed by a particular drug treatment, learning to discriminate food objects is also delayed. There may well be a transition from a young brain that can imprint to a somewhat older brain that can learn to identify and direct pecking towards food objects.

Likewise, there may well be an associated transition from a phase of imprinting to a phase of increased fear of novelty. However, here we mean fear of novel objects akin to the imprinting stimulus, not small novel objects that can be pecked at and ingested.

There are sensitive phases of development. The degree to which they are interrelated is not yet known. Layered upon them are other cycles, such as the diurnal cycles of sleep and learning performance, which further complicate matters. Unravelling these complex interactive processes is the challenge of understanding behavioural development.

4

BRAIN DEVELOPMENT AFTER HATCHING

Summary

- Unilateral treatment of the left or right forebrain hemispheres of the chick with cycloheximide or glutamate reveals asymmetries and a different developmental time-course for each hemisphere.
- Glutamate and γ-aminobutyric acid (GABA) are two amino acids which act as neurotransmitters and play important roles in early brain development (neural plasticity).
- Receptors for glutamate (NMDA-type) are elevated in the left intermediate medial hyperstriatum ventrale (IMHV) following imprinting or passive avoidance training involving pecking at a bitter-tasting bead.
- The chick is being used as a model to understand the cellular and molecular correlates of memory formation, and the present state of knowledge in this field is covered.
- Communication between the left and right sides of the brain is essential, and there is evidence that it becomes fully established over the first weeks posthatching.
- Synapse formation in the forebrain is largely completed before hatching, but the synapses do not begin to take on their mature form until three weeks after hatching, and the maturation process continues until the chick is 10 weeks old.

Introduction

Chapters 1 and 2 discussed the development of brain and behaviour before hatching and the influence of various forms of environmental stimulation on this development. In Chapter 3 attention was turned to early learning in the posthatching period and the need to understand the sensitive periods for learning. The projected aim was to place brain development in an ethological setting, and thus highlight its interaction with environmental influences. For example, although it was thought that the onset of the sensitive period for imprinting was determined by the stage of development of the brain, it is now clear that this is not exactly so. Studies of imprinting in the laboratory have shown that it is perfectly possible to visually imprint chicks within hours of hatching but this is not the stage of development at which it would normally occur in chicks incubated and raised by a hen. In chicks raised by the hen, visual imprinting to the visual characteristics of the hen is delayed until the motor abilities of the chicks have matured enough to allow them to fully stand and walk and to begin to scan the visual environment. The brain might be capable of forming imprinting memories of visual stimuli even well before hatching, but it does not do so; that is, the neural mechanisms of forming imprinting memories are present well before the onset of the sensitive period for visual imprinting. Of course, the development of the motor mechanisms used in standing and walking requires neuromuscular integration and involves brain mechanisms, but essentially the state of development of the neural circuits involved in imprinting learning and memory formation *per se* do not dictate when the sensitive period for imprinting will begin.

The end of the sensitive period for imprinting may also be dependent on experience, rather than simply being a phase of brain maturation. That is, the behavioural act of imprinting might bring about changes in those neural circuits used for imprinting, as will be discussed in this chapter, and in so doing cause the ending of the sensitive period for imprinting.

Thus, brain development cannot be seen to effect behaviour by a unitary process of causality from brain to behaviour, following a programme for development determined by the genes. Both brain and behaviour change as manifestations of development that occur in correlation with each other. Just as our thinking about behaviour has moved away from the idea of a division between instinct and learning, so too our thinking about brain mechanisms and behaviour has moved away from the concept of genes and environment as separate influencing factors. It is not possible to discuss brain development in the absence of behavioural development, and it might follow from this that brain development should not be studied without simultaneously investigating behavioural development.

Nevertheless, many of the studies to be cited in this chapter have focused on brain anatomy, neurochemistry, etc. without conducting behavioural measurements. Some researchers taking this approach attempt to compensate for this lack by citing behavioural studies conducted by other researchers (for example, by linking known sensitive periods in behavioural development with the observed changes in brain anatomy or neurochemistry) but, as Chapter 3 showed, the developmental stage at which sensitive periods occur can be markedly influenced by incubation and rearing conditions as well as the strain of chick. Therefore, inaccuracies are inevitable unless brain and behaviour are studied together within the same experimental paradigm. It is impossible to conduct a comprehensive investigation of the neural correlates of memory formation without including measures of behaviour. Many researchers have done just this and much of the important information about brain development in the early posthatching period has come from the use of the chick as a model in which to study memory formation.

Asymmetrical Development of the Forebrain Hemispheres

Although most of the main aspects of neuronal development have occurred before hatching, the chick's brain continues to develop for some weeks after hatching and to respond to environmental influences in terms of changed subcellular structure and neurochemistry.

Lateralization revealed by cycloheximide treatment

Some of the developmental changes are lateralized to one hemisphere or the other according to the age of the chick, and follow a particular time-course. This aspect of posthatching development has been revealed by injecting cycloheximide, which inhibits protein synthesis, into either one of the hemispheres of chicks of different ages, followed by assessing the outcome on the behaviour of the chick (Rogers, 1986, 1991). The injection of cyclohex-imide has no marked acute effects apart from increasing drowsiness for a brief time (Rogers *et al.*, 1977), and its longer-term effects are not at all obvious in the home cage. They are revealed by specialized tests. As mentioned in Chapter 3, if the drug is injected into the left hemisphere on day 2 posthatching, in the second week of life the chick is unable to learn to discriminate or categorize grain from pebbles within the 60 pecks allowed in the pebble-floor task depicted in Fig. 3.4. Similar treatment of the right hemisphere on day 2 has no effect on the chick's ability to discriminate grain from pebbles (Rogers and Anson, 1979). This clearly demonstrates lateraliza-tion of the hemispheres, but the result varies with the age and sex of the chick.

For females, a single injection into the left hemisphere on day 2 impairs learning to categorize grain from pebbles, but injection at one time before day 2 or from day 4 until day 11 has no effect. Nor is there an effect of treating the right hemisphere over this time span (Rogers, 1991, and Fig. 4.1). On day 12 both hemispheres become susceptible to cycloheximide treatment; injec-tion of either the left or right hemisphere impairs learning on the pebble floor to some extent. Then, on day 13 there is again no effect.

In males, learning on the pebble floor is impaired by an injection given to the left hemisphere at any time in the first week of life, whereas a matched injection of the right hemisphere has no effect (Rogers and Ehrlich, 1983; and Fig. 4.1). On day 7 neither hemisphere is susceptible to the treatment but on day 8 the left hemisphere becomes briefly susceptible only to lose this effect by day 9 and thereafter. The right hemisphere has a brief period of susceptibility on days 10 and 11, following that of the left hemisphere. These patterns of

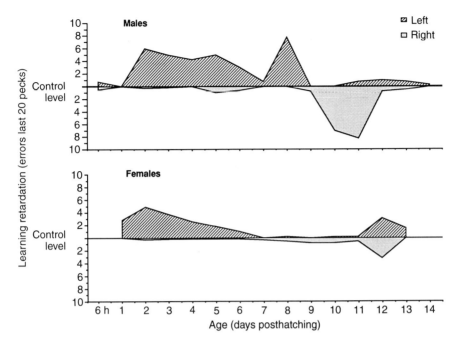

Fig. 4.1. Lateralized changes during development of the left and right hemispheres of male and female chicks. A single injection of 20 g of cycloheximide in 25 μl of 0.9% saline was injected into either the left or right hemisphere of chicks at various ages over the first two weeks of life posthatching. Each chick was injected once only and then on day 14 tested on the pebble floor task (see text). The number of pecks at pebbles in the last 20 pecks of the 60 pecks allowed in the task is plotted relative to the same value obtained from matched, control chicks treated with saline only at the equivalent age and also tested on day 14. The scores for chicks treated with cycloheximide in the left hemisphere are plotted above the control line (hatched area) and those for chicks treated in the right hemisphere are plotted below this line (dotted area). Higher scores either above or below the line indicate more errors and, therefore, slower learning relative to the controls. Controls shift from pecking randomly at grain and pebbles to pecking almost exclusively at grain in the last 20 pecks. When administered at certain ages the cycloheximide treatment prevents this shift away from random pecking. Note the differing ages at which cycloheximide affects the left hemisphere compared to the right and the male/female differences. Adapted from Rogers (1991).

drug action demonstrate that each hemisphere follows its own rather independent time-course of development with sharp changes in susceptibility to this particular treatment.

Lateralization revealed by glutamate treatment

Hambley and Rogers (1979) have shown that cycloheximide affects learning rate by raising the levels of glutamate and aspartate in the amino acid pools of the brain as a consequence of inhibiting protein synthesis. Therefore, injection of glutamate or aspartate alone has the same effect on learning as does cycloheximide (Howard *et al.*, 1980; Rogers and Hambley, 1982). At high doses glutamate is known to be neurotoxic in chicks (Snapir *et al.*, 1973) and also in mammals (Olney, 1978). At the lower doses effective in producing slowed learning on pebble floor learning (0.10 μmol per hemisphere is effective; Hambley and Rogers, 1979; Rogers, 1982a) glutamate is unlikely to destroy neurons but rather to affect their development by acting as an excitatory neurotransmitter. It possibly promotes the formation of inappropriate connections between neurons so that the brain becomes less efficient in learning.

It is known that iontophoretic application of glutamate to neurons facilitates synaptic modification in response to environmental stimulation (Greuel *et al.*, 1988). A similar interaction between glutamate stimulation and environmental stimulation appears to occur in the young chick brains treated with glutamate, because the glutamate is effective in changing behaviour only if the chick receives certain visual stimulation during the period for which the cycloheximide or glutamate is active (Rogers and Drennen, 1978; Sdraulig *et al.*, 1980). If the chick is injected and placed immediately into a white bucket with diffuse illumination for 3 h, the treatment has no effect. Similarly, if the chick views either stationary or moving parallel stripes, the treatment has no effect. By contrast, learning is impaired if the chick views intersecting lines for 3 h following the drug treatment. These results indicate that glutamate may be acting on the higher-order visual neurons in the forebrain which respond to angles. It is probable that, as in cats, during the early posthatching stages of visual experience neuronal connectivity at this level of organization is being modified in response to visual experience, and the glutamate treatment disrupts this process to leave long-lasting effects on the processing of visual information. Consistent with this hypothesis, there is evidence that glutamate is used in visual pathways in vertebrate species, including the chick (Voukelatou *et al.*, 1992).

These data for the chick indicate the existence of mechanisms similar to the modification of ocular dominance columns (Wiesel and Hubel, 1963) and orientation selective units (Blakemore and Cooper, 1970) in the kitten visual cortex. During sensitive periods soon after eye opening, the ocular dominance columns of the visual cortex of the cat are modified by monocularity or strabismus, and the orientation selective units in the visual cortex are modified by experience of lines oriented in one direction only. Glutamate-sensitive N-methyl-D-aspartate (NMDA) receptors appear to be involved in these plastic changes of the visual cortical neurons, since systemic application of ketamine, a non-competitive antagonist of NMDA receptors, with xylazine after monocular exposure prevents the shift in ocular dominance columns (Rauschecker and Hahn, 1987).

Thus, and perhaps not surprisingly, in both birds and mammals it seems that glutamate receptors, particularly of the NMDA type, play an essential role in the differentiation of neurons in the brain. It must, of course, be recognized that so far no-one has used electrophysiological recording of visual units in the chick forebrain to assess the effects of glutamate agonists or antagonists on neuronal specificity for visual stimulation, but this is a distinct possibility for future research. On the other hand, lateralized neurochemical events during development, such as those found in the chicken, might well be explored in mammalian species. We now know that functional lateralization of the cerebral hemispheres occurs in many species of mammal (Bradshaw and Rogers, 1992), and now it would be most interesting to explore the development of that lateralization.

The experiments using injection of glutamate into either the left or right hemisphere of the forebrain of the chick have revealed developmental changes in lateralization, but so far there has been no attempt to localize the site (or sites) at which the cycloheximide or glutamate may be acting. Multiple sites of action may be indicated by the fact that the treatment affects copulation and attack behaviour as well as performance on the pebble floor. Treatment of the left hemisphere on day 2 posthatching leads to an elevation of copulation and attack, as scored in standard tests using the hand as a simulated chick (Howard *et al.*, 1980; Bullock and Rogers, 1986). It is just as if the chick has been injected with testosterone (Chapter 5). Treatment of the right hemisphere has no such effects. In addition, auditory habituation is slowed after injection of cycloheximide or glutamate in the left, but not the right, hemisphere (Rogers and Anson, 1979; Howard *et al.*, 1980). All of these effects need to be assessed following more localized application of glutamate, or one of its more potent analogues (e.g. kainic acid or ibotenic acid) into particular sites of the forebrain.

It is possible that the lateralized effects resulting from the unihemispheric treatment at different ages reflect lateralized changes in receptor numbers for glutamate. For example, a site with a greater number of receptors for glutamate may be more susceptible to the application of glutamate to that hemisphere. Alternatively, the lateralized effect of glutamate may result from a difference in the affinity of glutamate receptors between the left and right hemispheres. Binding of tritiated MK-801, a NMDA antagonist which attaches itself inside the ion channel of the NMDA-type glutamate receptor (Foster and Wong, 1987), is twofold higher in the right hemisphere of two-day-old chicks incubated in darkness and held in darkness until sampling at day 2 (Johnston *et al.*, 1994a). Exposure of the embryo's right eye to light leads to a significantly higher level of binding in the left hemisphere in one-day-old chicks, and exposure of the left eye to light reverses the direction of asymmetry to give a higher level in the right hemisphere (Fig. 4.2). Therefore, binding to glutamate receptors is elevated in the hemisphere receiving visual inputs. Visual imprinting after hatching also generates asymmetry by elevating the binding levels for glutamate receptors in the left hemisphere

(Johnston *et al.*, 1993), as will be discussed fully in the next section. These binding data indicate that the susceptibility of the left hemisphere in the first week posthatching may depend on the greater number and/or affinity of glutamate receptors.

Lateralization of GABAergic neural connections

There are also hemispheric differences in binding to receptors for other neurotransmitters, and these levels are also influenced by visual experience.

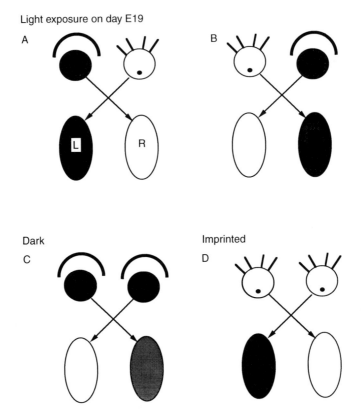

Fig. 4.2. The effect of light stimulation on day E19 of incubation on binding of ligands to NMDA-type glutamate receptors in the left or right hemispheres of the one-day-old chick (see text for details). The cartoons are as in Fig. 3.10. Black indicates a high binding level in the particular hemisphere, grey an intermediate level and white a low level. Intensity of shading indicates the magnitude of left–right hemisphere difference *within* a treatment group only. The binding to both of these types of receptors is higher in the hemisphere contralateral to the eye which has received exposure to light from day E19 to hatching. A represents the normal condition, B reversed asymmetry following manipulation of the embryo and exposing the left eye to light while the right eye is occluded, C indicates the asymmetry present in dark incubated chicks, and D the marked asymmetry following imprinting.

In one-day-old chicks that have received exposure of the right eye to light (around 120 lux) from day E19 of incubation, binding by muscimol to γ-aminobutyric acid (GABA$_A$) receptors (Chapter 1) is significantly greater in the left hemisphere than in the right, and in chicks that have received exposure of the left eye to light prior to hatching this asymmetry is no longer present (Johnston *et al.*, 1994a). In fact, in the latter chicks there is a trend towards greater binding in the right hemisphere. Hence, the hemisphere receiving visual stimulation develops a greater number of GABA receptors (c.f. Fig. 4.2). These must be primarily GABA$_A$ receptors because muscimol binds mainly to this subtype of GABA receptor. As GABA receptors are known to be important in a wide range of behaviours, including learning and memory, this light-dependent asymmetry might have wide-ranging consequences on behaviour (for a review of GABA and behaviour see Paredes and Ågmo, 1992).

Stewart and Bourne (1986) have also measured tritiated muscimol binding to GABA receptors in chicks and embryos exposed to 12 h light/12 h dark throughout incubation. The measurements were made at various ages between day E12 to day 21 posthatching. They reported an absence of any significant difference between binding in the left and right hemispheres. However, this was tested by an overall analysis of the total data set across all ages, which appears to have masked a significant asymmetry in one-day-old chicks. Judging by the figure in their paper, there is greater binding to GABA receptors in the left hemisphere, with no overlap of the standard errors around the mean values for each hemisphere. As the normal orientation of the embryo allows light stimulation of the right eye, this result is consistent with that of Johnston *et al.* (1994a), showing that monocular light stimulation of the embryo elevates GABA receptor binding in the contralateral hemisphere. At other ages prior to and after hatching there may indeed be no asymmetry. In fact, it appears that GABA binding reaches a peak in the left hemisphere on the first day after hatching and declines thereafter. In the right hemisphere it declines from day E18 on (see Fig. 1b in Stewart and Bourne, 1986).

Other researchers (Fiszer de Plazas *et al.*, 1991) have demonstrated a clear effect of viewing patterned visual stimuli on the development of GABA receptor sites in the optic tectum of chicks in their first week of life posthatching. By the end of the first week of exposure to black and white squares (patterned light) there is an elevated GABA receptor number relative to the level present in chicks exposed to plain white walls. Although the chicks exposed to the pattern might have visually imprinted on it, with consequent changes in receptor binding in the forebrain (see next section), it is most probable that the changes in receptor number in the optic tectum are due to stimulation *per se* and not learning.

Rearing chicks in continuous light raised GABA receptor number in the optic tectum by some 17% above the level for chicks reared under a 12 h light/12 h dark cycle (Gravielle *et al.*, 1992). As a result of the changed GABA receptors in the optic tectum, tectofugal visual inputs to the forebrain would be changed, because GABA is an inhibitory neurotransmitter that modulates local circuits within the optic tectum. The number of GABA receptors in the optic tectum reaches a peak at the end of the first week posthatching. Thereafter, the number of GABA receptors declines to stabilize at the lower levels characteristic of adults (Batuecas *et al.*, 1987; Rios *et al.*, 1987). Given that the increase over the first week of life posthatching depends on rather specific visual stimulation, this suggests a high level of neural plasticity in the optic tectum extending until the end of the first week of life (Fiszer de Plazas *et al.*, 1991). Despite the fact that the visual system of the chick is well developed by hatching, it is modified, even at the level of the optic tectum, by visual experience in the early posthatching period. Indeed, experience-dependent changes in GABA receptor number occur at many levels of organization during periods immediately before and after hatching.

Cellular and Molecular Correlates of Memory Formation

Many of the most marked subcellular changes that occur in the early posthatching period are related to memory formation. After early learning a number of neurochemical and subcellular changes in structure take place in the telencephalon. Molecular correlates of memory formation can be studied with relative ease in chicks because they form powerful and stable memories in early life (Andrew, 1991a). Also, a precise, brief learning period has advantages in experiments which attempt to link learning and memory to molecular changes. Indeed, it is the study of molecular correlates of memory formation in the young chick which has considerably advanced our understanding of this area in recent years.

Imprinting

As imprinting is a powerful form of learning in the chick, it has been selected for investigation of the neural bases of memory formation and recognition (Horn, 1990). The earliest investigations of the neurobiology of imprinting demonstrated a significant increase in the incorporation of radioactive lysine into protein and radioactive uracil incorporation into ribonucleic acid (RNA) molecules in the roof of the forebrain following visual imprinting training with the chicks in the wheel illustrated in Fig. 3.2 (Bateson *et al.*, 1972). The radioactive lysine or uracil was injected into the chick prior to its being imprinted and, following a period of exposure to the imprinting stimulus, various regions of the brain were assayed for incorporation of the molecules into protein or RNA, respectively. The increased incorporation was localized to the forebrain roof, and did not occur in the base of the forebrain or the midbrain.

Next it was necessary to prove that the increased incorporation in the forebrain roof occurred as a correlate of learning itself, rather than simply being the result of exposure to the visual stimulus without learning. This was achieved by exposing the chicks to the imprinting stimulus for varying periods of time on day 1 (20, 60, 120 and 240 min) followed on day 2 by measuring incorporation of radioactive uracil into RNA as a consequence of a fixed period (60 min) of re-exposure to the imprinting stimulus (Bateson *et al.*, 1973). Less exposure on day 1 should lead to more learning on day 2 and therefore incorporation of a greater amount of uracil into RNA. This was, in fact, the result obtained: the incorporation of radioactive uracil into RNA on day 2 was negatively correlated with length of training on day 1. As the chicks in each group received the same period of exposure to the imprinting stimulus on day 2 following the injection of radioactive uracil, the varying levels of incorporation reflected the amount of learning performed on day 1 and not simply exposure to the imprinting stimulus as occurred on day 2. A further experiment (Bateson *et al.*, 1975) demonstrated that the amount of incorporation of radioactive uracil into RNA correlated with the strength of preference for the imprinting stimulus (i.e. percentage imprinting) scored in the choice test as in Fig. 3.2. That is, the chicks which learnt more during imprinting training synthesized more RNA, in the roof of the forebrain.

The intermediate medial hyperstriatum ventrale (IMHV)

Using autoradiographic techniques, the exact region of the roof of the forebrain in which these biochemical changes occur was narrowed down to the intermediate medial hyperstriatum ventrale (IMHV) (Figs 4.3 and 4.4). Figure 1.6 shows that there is increased uptake of 2-DG in the IMHV regions of each hemisphere. Kohsaka *et al.* (1979) found that imprinting caused increased 2-DG uptake in IMHV and also in the lateral neostriatum. A similar autoradiographic image is obtained from chicks injected with radioactive uracil and imprinted (Horn *et al.*, 1979). Increased incorporation of the radioactivity occurs in the IMHV. Thus, the IMHV is a brain region essential for imprinting memory formation. It is not involved in the emerging of the predisposition for the hen (see Chapter 3), because lesioning the IMHV regions has no effect on this preference (Johnson and Horn, 1986).

Imprinting has been shown to correlate with a number of structural and neurochemical changes in the IMHV region of the forebrain, and many of these changes are asymmetrical (Horn, 1985, 1990, 1991). Lesioning studies have revealed that the IMHV region of the left hemisphere is involved in both the early and later phases of storage of imprinting memory, but the IMHV region of the right hemisphere is involved in the early events of memory storage only (Cipolla-Neto *et al.*, 1982). If a chick is imprinted on day 1 and then the right IMHV is lesioned 3 h later, followed by lesioning of the left IMHV some 26 h later, no memory is present on retest (Fig. 4.5). By contrast,

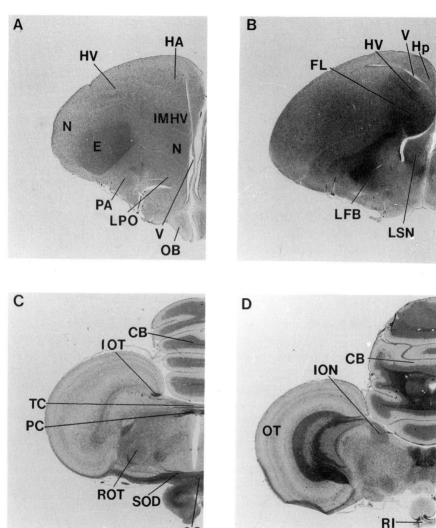

Fig. 4.3. A series of transverse sections through the left side of a two-day-old chick brain showing some of the main structures mentioned in the text. Sections are: A, through the middle of the forebrain; B, through the dorsal region of the forebrain and the thalamus; C, at a level approximately one third of the way through the optic tectum; D, further caudal but still within the region of the optic tectum. HA, hyperstriatum accessorium; HV, hyperstriatum ventrale; IMHV, intermediate medial hyperstriatum ventrale; N, neostriatum; FL, Field L; E, ectostriatum; LPO, lobus parolfactorius; PA, paleostriatum; OB, olfactory bulb; Hp, hippocampus; V, ventricle; LSN, lateral septal nucleus; LFB, lateral forebrain bundle; CB, cerebellum; OT, optic tectum; OC, optic chiasma; IOT, isthmo-optic tract; TC, tectal commissure; PC, posterior commissure; ROT, nucleus rotundus; SOD, supra-optic decussation; ION, isthmo-optic nucleus; RI, infundibular recess.

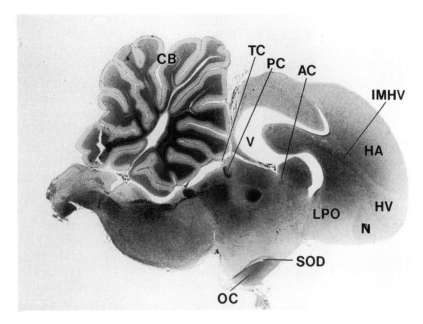

Fig. 4.4. Sagittal section taken from close to the middle of a brain from a two-day-old chick, stained with cresyl violet and luxol fast blue. It illustrates the commissures (tectal commissure, TC; posterior commissure, PC; anterior commissure, AC) and the supra-optic decussation (SOD) situated just over the optic chiasma (OC). Other structures are: CB, cerebellum; V, ventricle; IMHV, intermedial medial hyperstriatum ventrale; HA, hyperstriatum accessorium; HV, hyperstriatum ventrale; N, neostriatum; LPO, lobus parolfactorius.

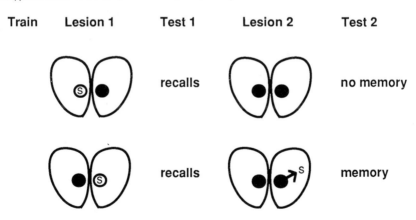

Fig. 4.5. Schematic representation of the experiment by Cipolla-Neto *et al.* (1982), which used sequential lesion of the left and right IMHV regions to reveal lateralized storage of imprinting memory. The chicks were trained and then lesioned in the right (top row) or left (bottom row) IMHV regions. At test 1 both groups could recall. Subsequent lesioning of the IMHV in the other hemisphere removed the memory store, S, in the left hemisphere but not the right (see text). Adapted from Horn (1985).

if the left IMHV is lesioned first and then the right, the chicks can recall the memory of the imprinting stimulus on retest (Horn, 1985, 1990). These results suggest that the left IMHV is involved in both the long-term and short-term processes of memory storage, whereas the right IMHV has a transient role in the initial (short-term) phase only (Fig. 4.5). In the right hemisphere, the memory appears to be shunted away from the IMHV to a store somewhere else in the hemisphere (Horn, 1985). This site is not yet known, although there is a suggestion that it might be the lobus parolfactorius (LPO) or the archistriatum (Rose and Csillag, 1985; Salzen et al., 1975).

Thus, the right IMHV appears to act as a buffer store for holding the memory before it is moved to another region that may have a larger capacity for storage and perhaps an ability to modify and extend the memory through subsequent experience. The latter would give the memory more flexibility for use in a variety of contexts (Horn, 1990), similar to the role in memory formation now assigned to the mammalian hippocampus (McNaughton et al., 1986). Horn and Johnson (1989) have suggested that, as is the case for the mammalian hippocampus, the right IMHV may add to the 'depth' of memory processing by contributing contextual information during learning. To exemplify this, the chick may first learn to recognize the hen as a stimulus in isolation, and even from a limited number of visual angles or with the hen in only one particular posture, and later it may add to this memory to allow recognition of the hen from all angles, in all of her postures and also in a wide variety of contexts. As 'depth' or complexity is added to the memory, it may be shifted to different neural circuits in different regions of the brain.

Lateralized involvement of the IMHV regions in imprinting learning is also apparent at the subcellular, electrophysiological and neurochemical levels. Neurons in the left IMHV respond to both visual and auditory stimuli of the type that are used in imprinting (Brown and Horn, 1992). After imprinting there are asymmetrical changes in the frequency of multiple-unit discharges recorded from the left and right IMHV regions (Davey and Horn, 1991). In dark-reared control chicks the left IMHV has a slight but significantly higher level of discharge. In imprinted chicks the asymmetry is lost immediately after training due to an increase in firing of the neurons in the right IMHV and a decrease in the left. By 6 h after training the asymmetry returns to an enhanced degree due to a fall of the firing rate in the right IMHV and an increase in the left. This increased rate of multiple-unit discharge in the left IMHV was confirmed by Bradford and McCabe (1992) at 20 h after imprinting. In addition, these researchers found a positive correlation between the behavioural preference for the imprinting stimulus and the mean firing rate of the neurons.

In contrast to the frequency of multiple-unit discharges, bursting activity is reduced in both left and right IMHV regions at 6 h after training. Interestingly, there is also a decrease in bursting activity in the HA region immediately after imprinting. This result is largely consistent with another mentioned in Chapter 1, namely that metabolic activity in the HA is reduced at hatching and remains so until sometime later in the first week of life posthatching (Rogers and Bell, 1989; Bell and Rogers, 1992; and see Chapter 1). Admittedly, the latter experiments found the onset of reduced activity in the HA to occur before hatching on day E20, whereas Davey and Horn (1991) found decreased discharge following imprinting on day 1 posthatching, but their chicks had been incubated and reared in darkness until they were imprinted. The embryos and chicks used to study metabolic activity by 2-DG uptake were exposed to light and therefore may have begun to visually imprint before hatching. The relative roles of the two visual systems in imprinting, as well as other sensory stimulated aspects of HA activity, deserve further attention. Either the occurrence of imprinting suppresses neural and metabolic activity in HA, or the reduction occurs in preparation for imprinting. Davey and Horn (1991) found that the reduced frequency of bursting activity in HA occurred immediately after the imprinting experience, followed a few hours later by a similar reduction in the left and right IMHV regions. It should also be noted that they used visual imprinting. Visual inputs reach IMHV indirectly via the HA, to which the thalamofugal visual system projects, and from the ectostriatum (Horn, 1985 and Chapter 1). Therefore, imprinting may lead to a reduction of thalamofugal visual input to the IMHV during the period of memory consolidation. The reduced activity in the HA may represent either a shutting off of neural processing or a limiting of processing to essential functions only. Whatever the explanation, the imprinting period is clearly coupled to changed activity in the region of the forebrain which receives input from the thalamofugal visual system.

Imprinting learning leads to changes in the structure of neuronal synapses in the left IMHV (Bradley *et al.*, 1981). No significant changes in synaptic structure occur in the right IMHV. In untrained, control chicks the length of the postsynaptic density of the synaptic apposition zones is greater in the right IMHV than in the left. The size of the synaptic apposition zone must reflect efficacy of neurotransmission at the synapse, although it is not known exactly how. Training removes this difference by increasing the length of the postsynaptic density in the left IMHV. The increase in postsynaptic density profile is confined to axospinous synapses in the left IMHV (Horn *et al.*, 1985), and there is some evidence that some axospinous synapses in the mammalian brain are associated with receptors for glutamate (see McCabe and Horn, 1991).

Indeed, imprinting has also been shown to correlate with a lateralized change in the number of receptors for glutamate in the IMHV region. At 7–8 h after imprinting there is a significant increase in the level of glutamate binding in the left IMHV, and not in the right. McCabe and Horn (1988) measured [^{3}H]glutamate binding in chicks imprinted on a rotating red box at 8 h after they had been trained by exposure to the imprinting stimulus for 140 min. Controls were held in the dark, and not exposed to the imprinting stimulus. They found significantly increased binding in the left IMHV of the imprinted chicks, compared to the controls. The result was subsequently confirmed by Johnston *et al.* (1993), who found that after imprinting not only was there an increase in the number of glutamate receptors in the left IMHV, but also the affinity of these receptors for glutamate increased. The change in number and affinity of the receptors combines to increase the effectiveness of glutamate neurotransmission.

NMDA-type glutamate receptors involved in imprinting

A particular subtype of the family of glutamate receptors, the NMDA receptor subtype, appears to be the main one involved because McCabe and Horn (1988) also demonstrated an increased level of NMDA-displaced glutamate binding in the left IMHV. Moreover, the amount of NMDA-binding correlated significantly with percentage imprinting scores. That is, the chicks which imprinted more strongly showed a greater increase in NMDA-receptor binding in the left IMHV, a result which strongly implicates the glutamate-sensitive NMDA receptors in imprinting memory formation. The role of NMDA receptors in imprinting has also been demonstrated by binding of tritiated MK-801 which binds to the ion channel of the receptor. Johnston *et al.* (1994b) have found increased MK-801 binding in the left IMHV following imprinting.

There is accumulating evidence that NMDA receptors are involved in neural plasticity in mammalian as well as avian species. This is an important topic in its own right, and the reader is referred to reviews by Rauschecker (1991), Cotman and Iversen (1987) and Rogers (1993). The NMDA receptors have a role in memory formation and in the neuroplastic changes which occur in early life. If these receptors are blocked by the administration of a drug, such as the anaesthetic ketamine, the neuroplastic changes in response to stimulation or deprivation are prevented (Rauschecker and Hahn, 1987). Conversely, if these receptors are stimulated at random in a generalized manner by injecting glutamate or one of its analogues the normal developmental changes may be disrupted (see later).

The elevation of NMDA receptor number in the left IMHV appears to be a manifestation of long-term memory formation in the left IMHV as it does not occur before 6 h after training (Horn and McCabe, 1990; McCabe and Horn, 1991). Also, there is no increase in glutamate receptor number in the right IMHV and, as demonstrated by the lesioning studies (see p. 129), the right IMHV apparently plays no part in long-term storage of imprinting memory.

The increase in length of the postsynaptic density (as measured in electron micrographic sections and therefore indicating an increased size of the synaptic apposition zone) in the left IMHV of imprinted chicks occurs well in advance of the changes in glutamate or NMDA receptor binding, at around 3 h after imprinting (Bradley *et al.*, 1981; Horn *et al.*, 1985). Although NMDA receptors are associated with the postsynaptic density (Fagg and Matus, 1984), the initial change in the latter does not appear to be directly related to the increase in available NMDA receptors. Possibly, new NMDA receptors are inserted into the postsynaptic thickenings after a delay period, although there is, as yet, no evidence that the increased number of receptors occurs in the same synapses which have undergone an increase in synaptic apposition length.

Some pharmacological experiments indicate a rather specific role for glutamate receptors in the IMHV. For example, Takamatsu and Tsukada (1985) found no effects on imprinting memory itself following injection into the left and right-IMHV of 6-hydroxydopamine or haloperidol, which both decreased the catecholamine contents of IMHV, although there were other side effects on behaviour. Injection of atropine and α-bungarotoxin similarly had no effect on imprinting, but kainic acid impaired imprinting at 3 h after injection. The dose of kainic acid used reduced the glutamate content of the IMHV, but this was not maximal until 24 h after the injection, by which time the effects of the drug on imprinting had passed. As injection of glutamate itself had the same transient effect as the kainic acid, the researchers concluded that the low dose of kainic acid used was affecting imprinting temporarily by exciting glutamatergic neurons.

Recent research has shown that the effect of administering glutamate on imprinting may be localized to the right hemisphere only (Johnston and Rogers, 1992; Rogers, 1994). Injection of glutamate into the right hemisphere at 1, 3 or 6 h after imprinting training has been found to block recall of the imprinting memory, whereas similar treatment of the left hemisphere has no effect. The rather low dose (0.5 μmol) used in these experiments was probably excitatory. The effect produced by treating the right hemisphere could be motivational, but this seems unlikely since the activity level of the chicks is unchanged. At 9 h after training Johnston found that glutamate treatment of either hemisphere is ineffective. Therefore, temporarily elevating glutamatergic mechanisms in the left hemisphere has no significant effect on the encoding of imprinting memory, whereas treatment of the right hemisphere temporarily prevents recall of memory from either hemisphere. Leakage of glutamate from the right hemisphere into the left cannot be the explanation for the result, because direct glutamate treatment of the left hemisphere is without effect on recall. Apparently, the learning experience leaves biochemical and structural traces in the left IMHV, and other forebrain sites in the right hemisphere are used for recall in the early stages of memory formation. The blocking of recall at this early stage after training caused by injecting glutamate into the right hemisphere is likely to involve a region outside of the right IMHV, since recall in binocularly tested chicks is not prevented by placing a lesion in the right hemisphere (Cipolla-Neto *et al.*, 1982). Further research using a variety of pharmacological agents injected into either hemisphere to impair the action of different neurotransmitters, together with thorough analysis of the effects on imprinting as opposed to side effects on behaviour, is needed before we can be certain of these results, but glutamate receptors do appear to be essential for imprinting memory formation or retrieval.

Expression of NMDA receptor activity is also important during training. If the right IMHV is lesioned and then an NMDA receptor antagonist, known as D-AP5, is injected into the left hemisphere prior to training, imprinting is significantly impaired (McCabe *et al.*, 1992). The lesion of the right IMHV makes the left IMHV play an essential role in imprinting and thus confines the NMDA receptor changes to the left hemisphere where the drug is acting.

Other neurotransmitters and molecular events involved in imprinting

Receptors other than the NMDA receptors are also involved in some forms of imprinting. For example, Davies *et al.* (1985) administered the noradrenergic neurotoxin N-(2-chloroethyl)-N-ethyl-2-bromobenzylamine hydrochloride (DSP4) to chicks prior to imprinting training with either the rotating box or stuffed hen. The treated chicks exposed to the box showed significantly reduced imprinting, but those exposed to the hen were unaffected. This implicates noradrenergic receptors in imprinting to some stimuli and not others. Apparently, noradrenergic receptors facilitate imprinting to artificial stimuli, or less complex stimuli. Consistent with the result obtained using DSP4, Davies *et al.* (1983) have shown that imprinting on the box causes a significant rise in noradrenaline levels in the chick forebrain, measured immediately after training and up to 50 h after training. Apparently, both NMDA and α-adrenergic receptors are involved in imprinting memory formation, a result which may not be too surprising since noradrenergic mechanisms have been implicated in neural plasticity in the kitten (Kasamatsu *et al.*, 1979; Bear and Singer, 1986).

Other researchers (Bradley and Horn, 1981; Haywood *et al.*, 1975; Longstaff and Rose, 1981) have suggested an involvement of cholinergic transmission in imprinting based on their findings of elevated activities of the enzymes used in synthesizing and breaking down acetylcholine and elevated QNB-binding to cholinergic receptor sites following imprinting. Hence, more than one neurotransmitter system appears to change after imprinting has occurred. The possibility of complex interrelationships between the various neurotransmitter systems in the IMHV following imprinting might well be recognized, but it would be difficult to investigate.

More recent studies have looked at the events immediately following imprinting and found elevated immunoreactivity to Fos protein in the IMHV (McCabe and Horn, 1993). Fos protein is produced by the immediate early gene c-*fos*, the expression of which marks metabolic activation of neurons (Sagar *et al.*, 1988). As will be seen later, expression of the c-*fos* gene also follows passive avoidance learning by young chicks and learning on the pebble floor by somewhat older chicks (Anokhin *et al.*, 1991). For imprinting, it is known that the increase in Fos protein correlates with the strength of preference for the imprinting stimulus (McCabe and Horn, 1993). It is a first step in the chain, or cascade, of neurochemical events and subcellular changes in structure that will follow such important learning in early life.

Other forebrain regions involved in imprinting

Most studies of the neurochemical aspects of imprinting conducted so far have concentrated on visual imprinting and the role of the IMHV region, but other regions of the forebrain may also be involved. For example, Johnston *et al.* (1993) found increased affinity of glutamate receptors in a region of the left hemisphere containing the lobus parolfactorius (LPO) and the archstriatum in imprinted chicks, compared to chicks held in darkness (Figs 4.3 and 4.4). Other regions of the forebrain might well be involved in chicks that have been imprinted to a combined auditory and visual stimulus or to an auditory stimulus alone. For example, re-exposure of 7-day-old guinea fowl chicks to their auditory imprinting stimulus (a tone) causes increased metabolic activity, as indicated by uptake of 2-DG, in three rostral brain regions, including areas of the hyperstriatum (labelled by the researchers as HAD), an auditory area called MNH encompassing part of the rostromedial neostriatum and hyperstriatum ventrale and LNH in the rostrolateral neostriatum and hyperstriatum ventrale (Maier and Scheich, 1987). A similar result was found for domestic chicks re-exposed to a combined visual and auditory imprinting stimulus (Wallhäusser and Scheich, 1987). It should be noted, however, that the latter researchers exposed the chicks to a stuffed hen that emitted the sound, and it was, therefore, a combined auditory and visual stimulus (see earlier). Therefore the 2-DG uptake may have occurred in both visual and auditory regions of the forebrain. In fact, HAD and LNH receive visual inputs, whereas MNH is activated by playing chick vocalizations to socially reared chicks (Wallhäusser and Schiech, 1987). It may be re-exposure to the auditory component of the imprinting that activates MNH.

In the re-exposed chicks, there was a 47% loss of dendritic spines on large type I neurons in MNH of the imprinted compared to control chicks. MNH receives its auditory input via the dorsocaudal neostriatum and it has reciprocal projections back to this latter region. It is those neurons which project from MNH to the dorsocaudal neostriatum which lose spines when imprinting occurs (Dörsam *et al.*, 1991), suggesting a change (possibly a decrease) in the reciprocal feedback system. Involvement of MNH in auditory imprinting is also indicated by the report of Braun *et al.* (1992) that administration of the NMDA receptor-antagonist APV into MNH blocks auditory imprinting.

It is possible that MNH plays a role in auditory but not visual imprinting, or at least combined auditory plus visual imprinting, but it seems likely that regions of the forebrain other than the IMHV may be involved in visual imprinting. In fact, Salzen *et al.* (1975) found that lesions of the lateral neostriatum impaired visual imprinting to objects suspended in the chick's home cage, whereas lesions which included the IMHV had no effect. Later, however, Salzen (1991) found evidence that IMHV lesions do impair imprinting to a moving patterned stimulus suspended in the home cage. Thus, task and stimulus type may influence the brain regions involved in imprinting memory formation. It will be recalled that Kohsaka *et al.* (1979) found that imprinting elevated the uptake of 2-DG in both the IMHV and the lateral neostriatum. The imprinting stimulus used by these researchers was a moving, red balloon, which leads one to suggest that movement may be the main aspect of the stimulus that leads to involvement of the lateral neostriatum.

Neural plasticity and the sensitive period for imprinting

As was discussed in Chapter 3, it was originally thought that, once the chick has imprinted, the sensitive period for imprinting closes and the chick will not imprint on another stimulus, or only with much greater difficulty (Bolhuis and Bateson, 1990). Depending on the stimuli used for imprinting this may be the case, but now it is known that imprinting, particularly to artificial objects, can be reversed (Bolhuis, 1991; and see Chapter 3). Even if the chick does imprint on a second stimulus, the imprinting to the first stimulus leaves a more permanent memory (Cherfas and Scott, 1981, and see Chapter 3). After a period of not being exposed to the second stimulus the chick reverts to a preference for the first imprinting stimulus. Generally speaking, therefore, during the first week of life the sensitive period for imprinting remains open until exposure to the appropriate stimulus occurs, and then it is closed. In other words, the neurochemical events which occur with imprinting may not only encode the memory, but also actively close the sensitive period for imprinting.

Little consideration has been given to the neurochemical events which may close the sensitive period. Some researchers argue that the sensitive period is closed by behavioural mechanisms, because the imprinted chick avoids strange objects and so it is no longer in a position to imprint even though it may still be neurobiologically capable of doing so. On the other hand, it is highly probable that there are neurochemical changes which prevent the imprinted brain from re-imprinting. They may be the same subcellular processes which encode the memory, or concomitant, but different, processes. The increase in glutamate receptor number and affinity which follows imprinting may, for example, close, or partially close, the sensitive period. However, it is not known for how long the glutamate receptor number and affinity is increased after imprinting. The maximum delay period between imprinting and assaying which has so far been tested is eight hours. It is possible that these increases are transient (McCabe and Horn, 1991), and are even followed by a decrease in glutamate receptor activity, the latter marking the end of the sensitive period for imprinting. In kittens, decrease in glutamate receptor number has been found to occur at the end of the sensitive period during which neural connections in the visual cortex are modified in response to the visual input received (Bode-Greuel and Singer, 1989). Between the second and fourth week of the kitten's life, glutamate receptor numbers increase dramatically and they remain elevated throughout the sensitive period for development of the ocular dominance columns in the cortex. Towards the end of the sensitive period, they decline to adult levels, suggesting that the sensitive period is closed by a fall in the number of active receptors for glutamate.

Recently completed experiments have shown that treatment of chicks on day 1 after hatching with a mixture of ketamine–xylazine or ketamine alone can extend the sensitive period for imprinting into the second week of life (Parsons and Rogers, 1992). The ketamine blocks NMDA receptors and the xylazine is an agonist of α_2-adrenergic receptors, which consequently reduces the release of noradrenaline (Starke *et al.*, 1989). In the first experiments the chicks were incubated, hatched and reared in complete darkness. Even under these conditions control chicks will not visually imprint in the second week of life, day 8 in these experiments. By contrast, chicks treated with ketamine–xylazine shortly after hatching were found to imprint well on day 8. In the retest on day 9, they showed about a 70% preference for the imprinting stimulus, compared to 50% in controls. In a later set of experiments it was found that raising the chicks in diffuse light did not affect the result: the chicks treated with ketamine–xylazine could still imprint on day 8, whereas controls could not. Somehow the blockage of NMDA receptors on day 1 posthatching renders the chick capable of imprinting, to a visual stimulus at least, in the second week of life even though it has some opportunity to imprint on visual and other stimuli during the first week of life. Presumably, the level of NMDA receptors has been modulated by the treatment.

If a change in NMDA receptor number does indeed mark the end of the sensitive period for neural plasticity and imprinting in normal chicks, it might also be the neurochemical basis for the shift from early forms of learning ability to later, more adult, forms of learning ability. For example, chicks in their first week of life show poor learning performance on the task requiring them to discriminate grain from a background of pebbles (Rogers, 1986). In the second week of life, their ability to perform this task is markedly improved. Of course, the improvement in this ability could depend on the maturation of an entirely different set of neurochemical processes, or it could be linked to the decline in neural plasticity for imprinting. There is some indication that the latter case may be true because the chicks treated with ketamine–xylazine can still imprint in the second week of life but they perform poorly on the pebble–grain (pebble-floor) discrimination or categorization task. Apparently, their brains have not made the transition into being able to learn to discriminate food from non-food (see Chapter 3).

The end of the sensitive period for imprinting phases into the beginning of a period in which plastic changes must correlate with other forms of learning. The neural basis of these later forms of learning in the chick has not yet been studied. It may differ either quantitatively or qualitatively from early learning, such as imprinting, which occurs during the sensitive period of maximal neural plasticity.

Passive avoidance learning

Another task being used to study the neurochemical correlates of early learning in the chick is the passive avoidance task involving pecking at a bead coated with methylanthranilate. This task is used in a number of laboratories studying the neurobiology of memory (Gibbs and Ng, 1977; Andrew, 1991a; Rose, 1991a,b). It is learned well by very young chicks, on the first day after hatching. The chick is trained by pecking at a bitter-tasting bead (usually a red bead) coated with methylanthranilate. This evokes a disgust response from the chick and it will avoid pecking a similar (red) bead presented at retest. It will, however, peck at a bead of another colour. Following passive avoidance learning, a cascade of cellular processes occurs spanning time-courses ranging from minutes to hours (Rose, 1991b) and in both the IMHV and LPO regions of the chick forebrain (Rose, 1989).

Preventing memory formation by injecting various drugs

The effects of injecting a wide variety of pharmacological agents into either the left or right hemisphere of chicks either prior to or after training on the passive avoidance task have been used to develop a model of memory formation; this work has been reviewed by Ng and Gibbs (1991) and Ng *et al.* (1992). In brief, the major findings of this group have led to the development of a model of memory processing with three distinct phases during which different types of drugs act to block recall, either temporarily or permanently. A short-term stage of the memory processing is formed by 5 min after training and decays by 10 min after training. This stage is susceptible to a wide range of agents that depolarize neurons. Such drugs include potassium chloride and glutamate. An intermediate stage of processing is said to fall between 20 and 50 min after training and it is characterized by disruption by drugs that block the sodium–potassium pump (e.g. ouabain or ethacrynic acid) or metabolic inhibitors such as dinitrophenol. A long-term stage of memory formation begins at 60 min after training and this can be disrupted by inhibitors of protein synthesis, such as cycloheximide. Formation of the long-term stage of memory formation can also be blocked by injecting 7-chloro-kynurenate, a highly selective antagonist of the glycine site on the NMDA receptor, into the

IMHV region prior to training (Steele and Stewart, 1993). This result illustrates the common dependence of imprinting and passive avoidance learning on NMDA receptors. Furthermore, this drug is as effective if it is injected into the left IMHV only, as it is when given bilaterally. It is ineffective when injected into the right IMHV. Its effectiveness in the left IMHV only might well be predicted from the known increase in the number of NMDA receptors in the left IMHV only following imprinting, but it contradicts another result that injection of glutamate blocks recall of imprinting when given to the right hemisphere and not the left (Johnston and Rogers, 1992; see p. 135). Of course, the different results could depend on different modes of drug action or on the tasks used in each case, but further experimental comparisons are needed.

The biochemical links, if they occur, between the short-term, inter-mediate and long-term stages of memory formation, remain speculative. Also, the exact timings vary according to social stress (e.g. whether the chicks are tested in pairs or isolation), which suggests an influence of hormones such as those of the pituitary–adrenal cortex axis (Gibbs *et al.*, 1991). In untreated chicks housed in pairs, there are two times at which memory recall is poor. These are at 15 and 55 min after training. Ng and Gibbs (1991) suggest that these times mark the transition between the stages of memory formation. In isolated chicks these times of poor recall occur after longer intervals from training. Administration of the hormones, adrenocorticotrophic hormone (ACTH) or corticosterone, delays the second time of poor recall, but does not affect the first time (at 15 min). The hormone vasopressin delays both the short and the longer times of poor recall, to 25 min and around 90 min respectively. Hormone levels (ACTH, vasopressin) may also influence the strength of the memory (Crowe *et al.*, 1990).

For speculation on the mechanisms underlying these effects see Gibbs (1991) and Gibbs *et al.* (1991). The model proposed associates the short-term stage of memory formation with electrical activity, presumably in reverberat-ing circuits, leading into dependence on the action of the sodium pump in the intermediate stage of memory processing. The sodium pump actively restores the resting levels of sodium and potassium ions across the neuronal membrane, which is particularly important following tetanic stimulation, and also actively contributes to the uptake of amino acids into the neuron. These amino acids might then be used to synthesize protein, and so lead to the long-term stage of memory.

The main problem posed by using drugs to study the biological correlates of memory formation is the specificity of any given drug, both at the cellular level and in terms of its effects on behaviour. For example, as discussed earlier in this chapter, cycloheximide does not only inhibit protein synthesis but it also elevates glutamate and aspartate levels in amino acid pools of the brain (Hambley and Rogers, 1979). Therefore, when cycloheximide causes amnesia of the passive avoidance, does it do so by preventing the formation of protein, as Gibbs, Ng and co-workers claim, or might it not do so by elevating glutamate and aspartate levels? Since Gibbs and Ng have found that injection of glutamate alone causes amnesia but only during the short-term stage, it would appear that the elevation of this amino acid is not the cause of the effect of cycloheximide on long-term memory. Nevertheless, is the long-term amnesia caused solely by the inhibition of protein synthesis or by its combination with elevated glutamate and aspartate levels? This has not been answered by pharmacological approaches to the problem.

Moreover, when a drug causes amnesia, it is important to determine its specificity on memory, rather than a side effect on motivation to respond or some other non-specific behavioural effect. In the passive avoidance task, for example, does pecking at the red bead in re-test mean that the chick has amnesia, or is there a non-specific enhancement of pecking at beads in general. There has been criticism of the earlier procedures used in the passive avoidance task on this account (Roberts, 1987). One way of determining the specificity of the memory is to present the chick with a blue bead as well as a red bead on re-test. Thus, the chick's discrimination ability can be tested. A chick that has formed a specific memory of the bitter-tasting red bead will avoid pecking at the red bead but will peck at the blue bead (Gibbs *et al.*, 1991). This is now the standard procedure used by the Gibbs and Ng group. Unfortunately, it does not completely solve all of the problems, as the chick attends to colour to a greater extent when it views a stimulus with its right eye (Chapter 3). Chicks that have been trained with a red bead coated in methylanthranilate show reduced pecking at both red and blue beads when tested using the left eye but, when they are tested using the right eye, they avoid the red but peck more at the blue bead (Andrew and Brennan, 1985; Andrew, 1988). That is, the chicks must encode colour as part of the memory of the task in the left hemisphere only, so that the colour of the bead is discriminated by the right eye only (Chapter 3, Table 3.1). Hence, in the binocular tests the discrimination between the red and the blue bead may be influenced by which monocular field the chick is using when it first catches sight of the bead, or by which hemisphere is dominant at a given age (Chapters 3 and 5). In fact, by placing beads simultaneously in both monocular fields of the chick, we have found evidence in support of this. At 30 min after training, the chick discriminates colour when it turns to peck the bead detected in the right visual field, but not when it turns to the left field (Rogers and Meurs, in preparation).

These results are mentioned here merely to emphasize the problems encountered when a seemingly simple learning task is used in conjunction with pharmacological agents to study the biological basis of memory formation. More sophisticated testing regimes are required to determine whether a given drug causes amnesia of a specific memory, or alters hemispheric dominance and therefore eye use in testing, and so on. Also, the same drugs need to be tested using a range of behavioural tests. It is, for example, known that both ouabain and ethacrynic acid cause a temporary amnesia following learning to discriminate grain from pebbles and after learning other appetitive tasks, but long-term memory still occurs (Rogers *et al.*, 1977). More comparison of the effects of the drugs on different tasks should help to solve such anomalies and better characterize each drug's effects.

There are fewer obstacles when the passive avoidance task is used to investigate memory formation by measuring the neurochemical and structural events subsequent to training with biochemical and histological techniques. This approach is not dogged by the problems of separating the side effects of the drugs from non-specific effects. The following section summarizes results obtained using this approach.

Cellular and subcellular correlates of passive avoidance learning

Coincident with passive avoidance training and immediately afterwards there is enhanced uptake of 2-deoxyglucose (2-DG) in the IMHV and LPO. In fact, as for imprinting, the left and right IMHV and LPO regions are differentially involved. Performance of the task results in different patterns of neural activity, detected by uptake of radioactive 2-DG in the left and right hemispheres (Rose and Csillag, 1985). The chicks were given a pulse of 2-DG lasting for 30 min just after training with the methylanthranilate-coated bead. Compared to controls, which pecked at a bead coated with water, the trained chicks showed increased accumulation of 2-DG in the IMHV and the LPO, the greatest changes occurring in the IMHV and the LPO of the left hemisphere.

The neurochemical changes which follow this event include first phosphorylation of the presynaptic membrane protein kinase C substrate, B50 (Burchuladze *et al.*, 1990), together with translocation of protein kinase C and genomic activation of the protein oncogenes c-*fos* and c-*jun* (Anokhin *et al.*, 1991). Somewhat later changes (1–6 h after training) include the synthesis of the protein, tubulin (Anokhin and Rose, 1991), synthesis of pre- and postsynaptic glycoproteins (Bullock *et al.*, 1990; McCabe and Rose, 1985) and 'bursting' of neuronal activity (Mason and Rose, 1987). Neuronal 'bursting' activity, recorded as extracellular multi-unit activity from the IMHV, lasts for up to 12 h after training (contrast this to the decline in bursting activity at 6 h after imprinting; Davey and Horn, 1991). Finally, at 12–24 h after training, morphological changes can be detected in synaptic structures. These include lengthening of the synaptic apposition zones and increased density of dendritic spines (Patel and Stewart, 1988; Stewart *et al.*, 1984). It will be remembered that auditory imprinting leads to a 47% decrease in the density of dendritic spines in MNH (p. 137). This opposite effect could be task or brain region specific. It is possible that the changes in the IMHV are part of a linked sequence or cascade (Rose, 1989, 1991b), but to prove this it is first necessary to determine whether all the changes occur in the same brain locations.

As for imprinting, the neurochemical changes and changes in subcellular structure which occur soon after passive avoidance learning are largely localized in the left hemisphere, although the timing of the events differs. Glutamate receptor binding is increased in the left IMHV and LPO, and not in the right. In fact, 30 min after passive avoidance training there is approximately a 40% increase in NMDA receptors in the left IMHV and LPO (Stewart *et al.*, 1992a). It will be noted that this increase in NMDA receptor binding following passive avoidance learning occurs much earlier than following imprinting (p. 134). Receptors for other neurotransmitters are also changed following passive avoidance learning and at different times after training. Binding of QNB to muscarinic receptors, α-bungarotoxin to nicotinic receptors, and serotonin to serotonergic receptors is varyingly altered at 30 min or 3 h after training, and muscimol binding to GABA$_A$ receptors is elevated at 24 h after training (Rose *et al.*, 1980; Aleksidze *et al.*, 1981; Bourne and Stewart, 1985). The latter is a long-term effect, quite distinct from the changes demonstrated for other neurotransmitters.

Although the biochemical studies have located changes in the left IMHV only, there is neurochemical evidence for a role of the right IMHV in the memory formation. Barber and Rose (1991) report that injection of 2-deoxy-D-galactose, which inhibits brain glycoprotein fucosylation, into the right hemisphere causes amnesia of the passive avoidance task, whereas injection of this drug into the left hemisphere has no effect on memory recall. This particular result is consistent with an earlier finding of a lateralized increased incorporation of fucose into glycoproteins in the right forebrain base (containing LPO) of trained chicks (McCabe and Rose, 1985).

Structural changes also occur in the LPO and the paleostriatum augmentatum (PA) regions following passive avoidance learning. At 24 h after training, the mean postsynaptic thickening length in the LPO region is greater on the right side in controls and on the left side in chicks trained in the passive avoidance task (Stewart *et al.*, 1987). Both sides of the LPO show a significant increase in the density of synapses. A more recent study by Hunter and Stewart (1991) found a significant increase in synaptic density (around 30%) in the left LPO at 24 h after training, and a lesser (10%) increase at 48 h after training. At 48 h after training the right LPO had an approximately 18% increase in synaptic density. These changes in synaptic density in the LPO regions are matched by increases in the number of dendritic branches and increased dendritic length (Lowndes *et al.*, 1991). In the PA region there is no change in the postsynaptic density thickening following passive avoidance training (Stewart *et al.*,1987), but the density of synapses and the number of vesicles per synaptic bouton is higher in the right PA of controls and this asymmetry is removed by training.

Although the passive avoidance task does not involve auditory learning, at least directly, there is some evidence that, as for auditory imprinting, the lateral neostriatum in the right hemisphere may have a role, since injection of glutamate into the right, but not the left, neostriatum causes amnesia (Patterson *et al.*, 1986). By contrast, injection of glutamate into the left, but not the right, IMHV causes amnesia. It should be noted that the spread of glutamate was not traced in this study; that is, although the IMHV and lateral neostriatum were targetted, other areas of the forebrain could have been affected. Nevertheless, different regions of each hemisphere appear to be used to store memory of the task, and possibly different forms of memory are laid down accordingly, as for imprinting. However, other evidence suggests a role of the left lateral neostriatum, as well as the left IMHV and LPO, in passive avoidance memory formation: at 30 min after training a significant 44% decrease in NMDA receptor binding occurs in the left lateral neostriatum, as opposed to the increase in left IMHV and LPO (Stewart *et al.*, 1992a).

Lesioning studies

Lesioning studies have demonstrated a differential role of the right and left IMHV regions in the laying down of memory of the passive avoidance learning task, as for imprinting (Patterson *et al.*, 1990). Chicks which have had the left IMHV lesioned on day 1 and are trained on day 2 are amnesic when tested for recall on day 3. A similar lesion of the right IMHV does not impair the memory formation. However, although lesioning the right IMHV does not result in amnesia in binocularly trained and tested chicks, it is capable of taking over from the left IMHV in chicks trained and tested monocularly. Sandi *et al.* (1993) found that in chicks wearing an eye-patch on the right eye (using the left eye), lesions of the right IMHV disrupt learning and memory acquisition, but lesions of the left IMHV are ineffective. These rather anomalous results, in comparison to the binocular situation at least, may occur as a consequence of the monocular experience causing shifts in the areas which are used for the memory. In addition, the same researchers have demonstrated that interocular transfer of the passive avoidance task occurs more readily from the right eye to the left (i.e. when the chicks are trained using the right eye and tested using the left) than vice versa; that is, the left eye has better access to information acquired by the right eye than the other way around. This result is consistent with that of Gaston (1984), who trained chicks on an operant visual discrimination task and found better transfer of learning from right eye use to left eye use than vice versa (see Chapter 3 for more discussion of Gaston's research).

Lesions placed in the hippocampus (a forebrain region equivalent to the mammalian hippocampus; Fig. 4.3) also affect recall of the passive avoidance learning. Sandi *et al.* (1992) placed bilateral lesions in the hippocampus at 18 h before training. When tested for recall at 3 h after training, the lesioned chicks were amnesic, whereas sham-lesioned ones had memory of the task. Lesioning of the hippocampus on the left side only produced the same degree of amnesia, whereas lesioning the hippocampus on the right side produced only a mild, and non-significant, degree of amnesia. This region of the left hemisphere, in addition to the IMHV, therefore plays a greater part in memory formation or recall of the tasks than does the equivalent region in the right hemisphere. As there are projections from the hippocampus to the IMHV (Bradley *et al.*, 1985), the effects of lesioning the hippocampus may be mediated by these projections. The hippocampus, however, seems to be involved only in the initial phases of memory formation because neither left-side nor bilateral lesions prevent recall when made 1 h after training. Although the hippocampus is known to be involved in spatial learning in other species (Sherry *et al.*, 1989; Chapter 6), this is not an obvious ability used in the passive avoidance task.

Comparing imprinting and passive avoidance learning

In an overall sense, the cellular events which correlate with imprinting and passive avoidance learning of the methylanthranilate bead task have much in common. Similar regions of the forebrain are involved, the changes are asymmetrical and similar neurochemical processes occur. Although there may be differences in the timing, it now appears that imprinting and passive avoidance learning rely on essentially the same neurobiological processes and these, in turn, closely parallel the processes which have been shown to underlie changes in neural connectivity which occur in the cortex of kittens during early development (Rauschecker, 1991). The latter is an example of neuroplasticity during development, but since both imprinting and passive avoidance learning are events of early life they may have more in common with neuroplasticity that do forms of learning that occur in later life. This remains to be seen.

Development of the Commissures and Decussations

We have discussed the behavioural and biochemical lateralizations that are present in the chick brain. It is clear that each hemisphere of the forebrain controls a different set of behaviours. This does not mean that each hemisphere operates completely independently of the other. Even though the avian brain differs from the mammalian brain by lacking the large corpus callosum, which connects the left and right hemispheres, it does possess a number of smaller commissures that connect the left and right sides. At mid-brain level there are the tectal (TC) and posterior (PC) commissures (Figs 4.3, 4.4. and 4.6). The tectal commissure connects the left and right tecta (layers 10 and 11) and the oculomotor nuclei on each side of the brain. The posterior commissure carries fibres from the pretectal nuclei, spiriform nuclei, isthmic nucleus and several other nuclei in the thalamus (Pearson, 1972) as well as fibres from the archistriatum (Phillips, 1966).

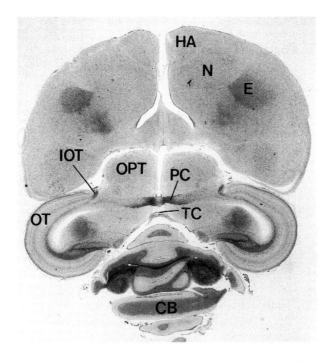

Fig. 4.6. A horizontal section through a two-day-old chick brain at the level of the tectal and posterior commissures (TC and PC). The connection of the left and right sides of the brain by these two commissures is shown. Abbreviations: HA, hyperstriatum accessorium; N, neostriatum; E, ectostriatum; IOT, isthmo-optic tract; OT, optic tectum; OPT, nuclei opticus principalis; CB, cerebellum.

Within the forebrain there is the anterior commissure (AC in Fig. 4.4) and the hippocampal commissure. These commissures connect homologous regions on either side of the brain. There are also decussations which connect non-homologous regions of the brain. For example, the supra-optic decussation (SOD; Figs 2.4, 4.3 and 4.4) consists of a bundle of axons which cross the midline as they travel from regions in the thalamus to the forebrain or vice versa (Ehrlich and Mark, 1984; Saleh and Ehrlich, 1984). It is in the SOD that the thalamofugal visual projections from one side of the thalamus to the contralateral hyperstriatal region of the forebrain cross the midline.

The function and development of these various left side to right side connections in the avian brain are of particular interest because it is known that the corpus callosum of the mammalian brain plays an essential role in maintaining lateralized function between the hemispheres (Denenberg, 1981). Via the commissural connections an area in one hemisphere is said to inhibit directly its homologous region in the other hemisphere from performing the same function. Thus, one hemisphere dominates the other (Chapter

6). It follows from this that lateralization is not fully developed until the corpus callosum is mature. Also, sectioning the corpus callosum should remove, or at least disrupt, the hemispheric lateralization.

The tectal and posterior commissures

This sectioning approach was adopted by Güntürkün and Bohringer (1987) to assess the role of the tectal and posterior commissures in lateralized function in the adult pigeon. They sectioned both of these commissures and examined the effects on monocular pecking performance. The pigeons were tested on a pattern discrimination task. Sham-operated control pigeons pecked faster when tested using the right eye, but pigeons with the commissures sectioned pecked faster when using the left eye. The lateralization was reversed following sectioning of the commissures. By some means, the direction of lateralization for control of pecking is determined by either one or both of these commissures. Güntürkün and Bohringer (1987) concluded that the lateralization depended on interaction between the optic tecta. This explanation is consistent with the finding of asymmetry in the size of the tectal neurons which receive projections which travel in the tectal commissure: cells in layer 13 of the tectum are larger on the right side (Güntürkün *et al.*, 1988; see Chapter 2 and also Chapter 6 for more detail). In other words, the neurons from the left tectum which send their axons in the tectal commissure and synapse on cells in layer 13 of the right tectum might be either greater in number than their counterparts going from right tectum to left tectum, or they may have more arborization and more end terminals. If so, the left tectum would have more influence on the right tectum than vice versa.

In the chicken, sectioning the tectal and posterior commissures on day 2 posthatching also affects lateralized pecking behaviour but in a different way (Parsons and Rogers, 1993). In control, unoperated and sham-lesioned chicks pecking at a small bead either habituated or showed no change in pecking over subsequent presentations. An unusual result was obtained with chicks that had the commissures lesioned: with each presentation of the bead they pecked more, and this response was significantly greater in those chicks tested using the right eye compared to the ones using the left eye. It was as if the right-eyed chicks recognized the bead but found it more interesting to peck at with each presentation, suggesting that they were unable to make a perfect match of the bead with the memory of it from previous presentations. A slight mismatch could lead to increased attention to the bead (Andrew, 1976) and hence increased pecking. Whatever the detailed mechanism involved, the results clearly indicate that the intact commissures suppress pecking elicited primarily by inputs provided by the right eye. It is known that there are inhibitory projections within the tectal and posterior commissures (Robert and Cuenod, 1969) and possibly those passing from the left optic tectum to the right serve to suppress pecking.

Sectioning the tectal and posterior commissures also unmasks lateraliza-tion of fear responses (Parsons and Rogers, 1993). With subsequent presenta-tion of the bead, fear responses decreased in all of the chicks tested, apart from those that had the commissures sectioned and were using the left eye. This indicates that fear is elicited by visual inputs to the right side of the brain, a finding which is consistent with that of Phillips and Youngren (1986) who found a greater involvement of the right archistriatum, compared to the left, in fear responding to a novel environment (Chapter 3). As the posterior com-missure carries axons projecting from the archistriatum, there is a distinct possibility that this is the brain region involved in fear responses to the bead. If so, lateralization at the forebrain level (archistriatum) must be suppressed by the intact commissure, because it is unmasked after sectioning the commissure. Therefore, unlike the corpus callosum of the mammalian brain, the tectal and posterior commissures do not generate lateralization, but suppress it. Maybe this function of the commissures relates to their presence at lower levels of integration, or maybe it is a commissural function unique to the avian brain. It is possible that the anterior commissure of the forebrain has functions similar to the mammalian corpus callosum, but this is not known.

The supra-optic decussation

Sectioning the supra-optic decussation also affects lateralization in the chick, but for different visually controlled behaviours (Rogers and Ehrlich, 1983). In addition to axons from a variety of neurons not driven by visual stimulation, the supra-optic decussation also contains the contralaterally projecting axons of the thalamofugal visual system. Since there are more contralateral projections from the OPT regions of the left side of the thalamus to the hyperstriatum on the right side of the forebrain than vice versa in chicks that have been exposed to light before hatching (Chapter 2), it is not surprising that the supra-optic decussation is involved in lateralization of those visual functions relying on input from the thalamofugal visual system. Discrimina-tion of grain from pebbles appears to be such a function, as it is reversed when the asymmetry of the thalamofugal visual projections is reversed; that is, when the left eye of the embryo is exposed to light, instead of the right (Chapters 2 and 3). The lateralization for pebble–grain discrimination is also affected by sectioning the supra-optic decussation (Rogers et al., 1986). After sectioning of the supra-optic decussation on day 2 posthatching, chicks tested in the second week of life show no lateralization of monocular performance on the pebble-floor task. In intact chicks and sham-lesioned chicks, perform-ance on this task is superior when the right eye is used. In the chicks with the supra-optic decussation sectioned performance is poor with either eye. In fact, performance is also poor when the lesioned chicks are tested monocularly, suggesting that the contralaterally projecting fibres are essential for discrim-inating grain from pebbles.

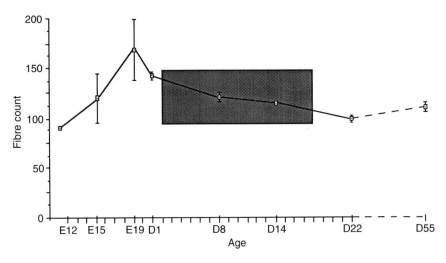

Fig. 4.7. The number of fibres in the supra-optic decussation ($\times 10^4$) in embryos (day E12–E19) and chicks posthatching (day D1–D55). The data are plotted as means with standard errors, apart from E12 and D14 as these data points are each for only one subject. Note that the number of fibres increases until day E19 and then declines over the first week or so posthatching. The shaded area indicates the ages over which there is asymmetry in the thalamofugal visual projections. Adapted from Rogers and Ehrlich, 1983.

Counting the number of axons in the supra-optic decussation of embryos and hatched chicks at various ages has revealed that the supra-optic decussation develops before hatching reaching a maximum fibre content around day E19 of incubation and thereafter, into the first week posthatching, it loses some 30% of its fibres to reach its adult size (Rogers and Ehrlich, 1983; Saleh and Ehrlich, 1984, and Fig. 4.7). It apparently reaches maturity at about the end of the first week of life, an event that correlates with some of the transitions in functional lateralization that occur at this age.

As discussed in Chapters 1 and 2, most of the neuron loss in other visual regions of the chick brain occurs prior to hatching. Although there is a decline in the number of fibres in the supra-optic decussation just prior to hatching the process continues well after hatching, which indicates a role for patterned or other more complex visual stimulation in the selective loss of the contralaterally projecting axons.

Chapter 2 discussed the role of light stimulation in the development of the thalamofugal visual projections and the way in which light stimulation of the right eye just prior to hatching causes asymmetry in the contralateral projections crossing the midline in the supra-optic decussation. Boxer and Stanford (1985) used horseradish peroxidase (HRP) labelling of neurons and counted a greater number of fibres projecting from the left side of the thalamus to the right hyperstriatum than vice versa. However, the technique of injecting fluorescent dyes into the hyperstriatum on each side of the forebrain to determine asymmetry, as used in my laboratory, does not permit determination of absolute numbers of projections. The ratios obtained by dividing a number of cells labelled contralaterally to an injection site by those labelled ipsilaterally to the same site (C/I ratio; Fig. 2.5) reveals asymmetry, but not whether that asymmetry results from a loss of contralateral projections from the right thalamus or a growth of contralateral projections from the left thalamus. The declining fibre numbers in the supra-optic decussation measured from day E19 to day 8 posthatching by Rogers and Ehrlich (1983) suggests that a loss is more likely, but this contradicts the finding of Boxer and Stanford (1985). Other evidence (Rogers and Bolden, 1991) also indicates that the asymmetry results from a growth of projections from the left side of the thalamus. Of course, both growth and loss of projections may occur simultaneously, and in opposite directions to produce lateralization. Also, it must be remembered that the thalamofugal visual projections make up only part of the supra-optic decussation and therefore the decline in fibre total number might be primarily due to fibres outside the thalamofugal visual system.

By the end of the third week of life posthatching there is no longer any significant asymmetry in the thalamofugal visual system. Male chicks injected with the fluorescent dyes on day 21 posthatching have no asymmetry, whereas those injected on either day 2 or day 12 have clear asymmetry (Rogers and Sink, 1988). In males, therefore, the asymmetry in the thalamofugal visual projections is transient. There are no comparable data available for females older than two days posthatching. Nor is it known whether the disappearance of the asymmetry, or rather the reappearance of symmetry, in males is due to a growth of projections from the right side of the thalamus to the left hyperstriatum or a loss of projections from the left side of the thalamus to the right hyperstriatum occurring between days 16 and 21 posthatching. As Fig. 4.7 illustrates, there is no change in the total fibre number in the supra-optic decussation over this period of development (Rogers and Ehrlich, 1983). This may simply indicate that any changes in the contralateral visual projections are not detected in the mean score of total fibre count, although this seems rather unlikely given the proportional representation of these particular neurons in the decussation (Saleh and Ehrlich, 1984) and the low variability of the scores between individuals in the posthatching period compared to those obtained in the prehatching period (Fig. 4.7). An alternative explanation may be that the loss of asymmetry in the C/I ratios of three-week-old chicks is caused by changes in the ipsilateral rather than the contralateral projections (for example, to a loss of visual projections from the left side of the thalamus to the left hyperstriatum). Only counting of, for example, HRP-labelled axons in the ipsilaterally and contralaterally projection tracts will answer this question.

Aside from this issue, it is interesting to reflect on the continued development of the thalamofugal visual projections well after hatching. We know that their early growth patterns are dependent on light stimulation of the embryo, with consequent development of the asymmetry (Chapter 2), but not whether their delayed development at the end of the second week of life posthatching is also dependent on light stimulation.

It is also interesting to speculate on the function of this asymmetry in the first two to three weeks posthatching. Is it merely an artefact of the orientation of the embryo in the egg, or does it play a critical role in early behaviour? The presence of asymmetry in these visual projections does not appear to be essential for visual imprinting after hatching. In fact, chicks hatched from eggs incubated and raised in darkness with only a brief period of priming in diffuse light imprint better than those exposed to light, particularly during and after hatching. At least, this is the case for visual imprinting in the laboratory setting. In a natural context, the embryos are likely to have been exposed to light, as have the chicks after hatching (Chapter 2). If the thalamofugal visual projections do have a lesser role in imprinting than do the tectofugal ones (p. 36), the absence of any relationship between imprinting and their asymmetry may be expected.

Chicks hatched from eggs in darkness also learn to discriminate pebbles from grain, when tested binocularly in the second week of life, as well as chicks exposed to light prior to hatching (Rogers and Krebs, in preparation). There is increasing evidence that the thalamofugal visual system is used to discriminate grain from pebbles, and that it requires a functional (intact) supra-optic decussation (Rogers *et al.*, 1986; p. 149). However, this form of visual discrimination learning does not require asymmetry in the thalamofugal projections. Further comparisons of chicks hatched from eggs that have been exposed to light (with asymmetry of the thalamofugal projections) with chicks hatched from eggs incubated in darkness (no asymmetry) may reveal differences in binocular visual performance, but so far the functional significance of this asymmetry in the intact chick eludes us, unless it relates to social behaviour in competitive situations, as discussed in Chapter 2.

One should be reminded that the asymmetries demonstrated by Güntürkün and colleagues in the tectofugal visual system are present in adult pigeons. Given that adult chickens look overhead using the left eye when they are played an alarm call (Evans *et al.*, 1993a; Chapter 3), one might postulate that a permanent asymmetry exists in the tectofugal system, as opposed to the thalamofugal system. To go a step further, it is interesting to consider the idea that lateralization in the visual system(s) may be connected to lateralization for auditory processing of species-specific vocalizations or other relevant auditory information. If so, the function of this particular form of lateralization may become apparent in tasks requiring auditory and visual integration.

Maturation of Synapses

So far we have discussed brain development over the first two or three weeks posthatching. Maturational changes continue to occur after this time. In particular, the synapses show maturational changes between three and 8–10 weeks posthatching. In chicks synaptic formation is temporarily separated from synaptic maturation (Rostas *et al.*, 1991). In an overall sense in the forebrain, proliferation of synapses occurs at a maximal rate on day E8 of embryonic development (Chapter 1), but it continues throughout the incubation period and until around the end of the second week posthatching (Rostas, 1991). Given that the thalamofugal visual projections continue to show developmental changes between days 14 and 21, it must be recognized that new synapses may be forming in at least some localized regions of the forebrain (in the hyperstriatum in this case) in even older chicks. There is evidence that catecholamine neurons are forming new synapses in the paleostriatum ventrale region of the two-week-old chick forebrain (Reynolds *et al.*, 1992). However, generally most regions of the forebrain have attained their full number of synapses long before this stage.

The synapses formed by this stage are not yet identical to those of the adult, however. Maturation of the synapses is required before they assume the appearance and neurochemical characteristics of synapses in the adult brain (Rostas, 1991). These changes occur in the postsynaptic density (PSD), a protein-rich specialization on the intracellular surface of the postsynaptic membrane. It is tightly bound to proteins in the postsynaptic membrane, which includes the receptors for neurotransmitters and ion channels. It is thought to regulate synaptic function, possibly by modifying receptors and ion channels. Changes in PSD may change receptor affinity and so enhance transmission at synapse. Even in adults the PSD is a dynamic structure, undergoing morphological and biochemical changes which may relate to change in synaptic activity (Siekevitz, 1985).

In newly hatched chicks the postsynaptic densities are much thinner than in adults (Rostas *et al.*, 1984). They also have lower levels of one of the major proteins in PSD, the α-subunit of calmodulin stimulated protein kinase II (α CMKII). From week 3 to about week 10 posthatching the PSD thickens and its level of α CMKII increases to reach adult levels (Rostas and Jeffrey, 1981). It is likely that this maturation is dependent on experience. Indeed, earlier and localized maturation of synapses may occur in response to specific learning, such as imprinting. Consistent with this hypothesis, it has been demonstrated that the expression of genes for α CMKII in the IMHV increases in ducklings that have been imprinted at 10–12 h posthatching and killed 2 h later, compared to ducklings kept in the dark (Kimura *et al.*, 1993). Presumably the same occurs with imprinting in the chick, and may relate to the known changes in synaptic apposition length that correlate with imprinting (p. 133); that is, maturation of synapses in the IMHV may occur as early as day 1 in imprinted chicks. If so, the overall delay in maturation of the majority of synapses in the brain until after the third week posthatching may merely reflect delayed learning in the majority of brain regions, or rather the accumulated effects of many learning experiences prior to that age.

The significance of a thicker PSD with more α CMKII is not yet known, but there is some indication that it reflects alterations in the synaptic response to stimulation, although whether it enhances or diminishes the response may vary in response to different forms of stimulation and in different regions of the brain (Rostas, 1991).

Much more research is required to understand the process of synaptic maturation, but it clearly demonstrates the prolonged nature of brain development in the posthatching chick. In the rat, synapse formation and maturation overlap in terms of the ages at which they occur, whereas in the chicken they are discrete events (Rostas, 1991; Heath *et al.*, 1992). Although proliferation of synapses occurs earlier in the chick, it appears that maturation of the synapses does not similarly occur at an earlier age. Indeed, the development of the chick may not be precocious in terms of synaptic maturation, although it is in all other known aspects of brain development. One might say that the precocious stages of development of the chick are associated with a brain that can form synapses earlier and in a ubiquitous manner throughout brain regions, compared to more slowly developing species. However, the subsequent maturation of those early-formed synapses is no different in precocial species.

Finally, given the important role of the NMDA-type glutamate receptors in early brain development (pp. 124–125 and 134–135), it should be stated that there is no change in MK-801 binding to NMDA receptor ion channels with maturation of the synapses in chick brain (Kavanagh *et al.*, 1991). Thus, maturation of PSD may be associated with neural plasticity, but is not linked to qualitative changes in the nature of the NMDA receptors themselves, either in number or affinity. Rather, it may be linked to more complex aspects of synaptic response efficacy, such as long-term potentiation. Of course, synaptic maturation may bring about changes in the affinity of receptors for other neurotransmitters, even though it does not do so for NMDA receptors.

Concluding Remarks

As this chapter has shown, the chick brain is an excellent model for studying the neurochemical and subcellular events that occur during development, as well as those that correlate with the formation of memory. Although memory processing is usually considered as separate from brain development or maturation, both processes have much in common neurochemically and behaviourally. Indeed, the brain does not develop without making memories, and the formation of memories influences the course of development. Both processes are, in fact, different aspects of neural plasticity, and both rely on modulation of NMDA receptors.

As imprinting and the passive avoidance bead task are forms of learning that occur in chicks soon after hatching, it could be argued that they may have more in common with neural plasticity during development than do other forms of learning occurring later in life. Learning in adult rats has, however, been shown to involve NMDA receptors (Melan *et al.*, 1991). It would be of much interest to study the neurobiological events that correlate with learning in older chicks. For example, filial imprinting might well be compared to sexual imprinting. The process involved in learning to discriminate grain from pebbles might also be compared to filial imprinting, since chicks do not perform the pebble-floor task well until the second week of life posthatching (Rogers, 1986).

It is already known that performance of the pebble-floor task leads to expression of immediate early genes (Anokhin *et al.*, 1991), as do imprinting and passive avoidance learning, but no other neurochemical changes that follow pebble-floor learning have yet been investigated. Nevertheless, since glutamate treatment of the forebrain leads to slower learning of the pebble-floor task, it is likely that glutamate receptors play a role in this form of learning. It would be most interesting to know whether the same regions of the forebrain (in particular, the IMHV and LPO) are involved in memory storage of the pebble-floor task as for imprinting and passive avoidance learning. The question to be addressed is whether all forms of learning by the chick involve the IMHV or LPO, or whether different tasks use different regions of the brain.

On the other hand, there is some evidence that different aspects of the memory of one task may be encoded in different regions of the brain. Patterson and Rose (1992) reported results suggesting that the IMHV stores memory of the colour parameters of the bead in passive avoidance learning, whereas the LPO stores other aspects of the learning task. They state that even simple associations are stored in the form of multiple and dispersed representations. Also, as Horn and Johnson (1989) have suggested, the memory may be shunted to different regions of the brain as 'depth' is added to it, for example, by adding contextual information or other associated information. Some memory systems may involve virtually all levels of processing in the brain, whereas others may be highly specific and localized to one site (Weiskrantz, 1990).

Moreover, different tasks may lead to the use of different regions of the brain, or even different neurochemical correlates. However, the latter seems rather unlikely as it is becoming apparent that the neural processes of memory formation, and also for neural plasticity, are similar in all species that have been studied in this way: namely, chicks, cats and rats (Rogers, 1994).

Age is also a variable that may influence the timing, if not the nature, of the neural events of memory formation. From the studies with chicks, it is known that lateralization of the hemispheres changes during development, and that early memory processing is strongly lateralized. Although both hemispheres may be involved in memory formation, they are involved differently. Different brain regions and time-courses of subcellular change occur in each hemisphere. In all of the imprinting studies discussed in this chapter, these lateralized events have been shown to occur in the brains of chicks incubated and hatched in darkness. Therefore, the chicks used would have lacked asymmetry of the thalamofugal visual projections. Following a more natural form of incubation in which the embryos are likely to receive sufficient light stimulation before hatching (Chapter 2), there may be even greater lateralization of the subcellular events correlating with memory formation.

The form of lateralization of memory processing may also vary with age, according to the lateralized development of the hemispheres. Unilateral injection of cycloheximide into the hemispheres at different ages over the first two weeks of life has revealed marked changes in lateralization. Thus, learning at different ages must be superimposed on different forms of lateralization. This may influence the lateralized events occurring after learning at a particular age. Would perhaps the neural events of passive avoidance learning or imprinting be lateralized in the opposite direction if they occurred on day 10 or 11 posthatching when the right hemisphere is temporally susceptible to cycloheximide? Is early learning qualitatively different from learning in later life because it occurs in a brain with different lateralization? These questions are important for understanding memory processes in all species. Not only have they arisen from studies of memory formation and development in the chick, but they can now be addressed by continued use of the chick as a model.

In all of these studies, there is a need for more detailed analysis of the chick's behaviour, not only to separate side effects of pharmacological agents from their genuine effects on memory, but to reveal more information about the process of memory formation. To illustrate, it would be worth investigating whether any form of behavioural bias might correlate with the neurochemical differences following imprinting. If, as Horn (1990) suggests, shunting the memory out of the right IMHV to other regions of the hemisphere is associated with adding contextual depth to the imprinting memory, left eye viewing should allow better recognition of the imprinting stimulus in a variety of contexts. Indeed, Vallortigara (1992b) has shown that chicks using the left eye recognize individual conspecifics from strangers, whereas those using the right eye do not (Chapter 3). In my opinion, attention to such lateralized responses in those paradigms used for detailed study of the biological basis of memory formation (imprinting or passive avoidance learning) should considerably enhance our understanding of it.

5

BEHAVIOURAL TRANSITIONS IN EARLY POSTHATCHING LIFE

Summary

- During the first two weeks of posthatching development the behaviour of the chick undergoes a number of sudden transitions or changes.
- The most marked and frequent transitions occur between days 8 and 12. For example, on day 8 chicks reared by the hen in a farmyard begin to fixate conspecifics and humans, on day 10 they display a peak in initiating movement away from the hen and moving out of her sight and on days 11 and 12 they being to perform social interactions such as frolicking.
- There are also sudden changes in fear behaviour and choice of familiar versus novel-coloured objects between days 7 and 10, and searching for food using spatial cues changes likewise over this period.
- Many of the transitions in behaviour can be linked to changes in hemispheric dominance, or dominance of the neural systems connected to one eye relative to the other.
- There are sex differences in hemispheric dominance, depending on age and context.
- Other changes in behaviour result from maturation of physiological control systems, such as the development of thermoregulatory ability and the systems involved in hunger and food searching.
- Transition into adulthood depends on hormonal changes and concomitant changes in agonistic and aggressive behaviour. Shifts in brain lateralization appear to be involved in these behavioural changes also.

Introduction

The previous two chapters have been concerned with the development of brain and behaviour in the early posthatching period. This chapter will focus on transitions in posthatching behaviour and accompanying changes in brain control mechanisms. In Chapter 3, the timing of sensitive periods for different kinds of learning were discussed. For example, the sensitive period for filial imprinting occurs early in the chick's life, the sensitive period for sexual imprinting occurs considerably later and learning to classify small objects into the categories of 'food' versus 'non-food' occurs in the second week posthatching. These examples illustrate that behavioural development passes through phases at which certain forms of learning are performed better than, and in preference to, other forms. At rather well-defined ages there are transitions from one learning state to another. Each learning state may in fact develop independently of any other and merely assume predominance at a particular age according to an apparent sequence as a result of its own discrete developmental programme. Alternatively, the transition from one state to the other may require closure of the earlier sensitive period before the second can occur. For example, as discussed in Chapter 4, the ability to perform well on the pebble-floor test may require closure of the sensitive period for filial imprinting. Thus, chicks treated with ketamine and xylazine on day two posthatching can imprint in the second week of life but their ability to discriminate grain from pebbles is poor, possibly because they have not made the transition from early learning capabilities to later learning ones.

The effects on pebble-floor performance of injecting cycloheximide into the left or right forebrain hemisphere at different ages throughout the first two weeks of life fully demonstrate marked transitions in development of the hemispheres that occur in the early posthatching period (Fig. 4.1). Day 2 marks the onset of the developmental events that are disrupted by the cycloheximide treatment, and these are occurring in the left hemisphere only. The development of the left hemisphere passes through various changes in susceptibility to cycloheximide between days 7 and 9, and the right hemisphere follows with transition into a susceptible period from days 9 to 12. One might therefore predict that between days 7 and 12 the chick would display marked transitions in behavioural states. This is the case, as will be discussed in this chapter. The focus here is on transitions in behavioural responses of any kind, not especially learning.

Transitions in General Behaviour

Some changes in behaviour occur rather gradually with increasing age, whereas others occur rapidly. The former may be marked by a transition period of only one day. Such a transition might well be overlooked if the chicks are not tested at least daily at the same time under controlled conditions. Transition periods of even shorter duration occur, but they are either not detected by daily testing or interpreted as diurnal cycles.

Without daily testing a transition point may well be missed if, for example, a given behaviour occurs on one day, not on the next but again on the following day, or vice versa. Thus, cycloheximide leads to slowed learning on the pebble floor when it is injected into the left hemisphere on day 8, but not on days 7 or 9 (Chapter 4, Fig. 4.1). In earlier testing paradigms this rapid transition was missed by failure to test on each of these days (Rogers and Ehrlich, 1983). By contrast, other transitions in which a new behaviour pattern emerges from a phase of either no performance or a very low level of performance of that particular behaviour are unlikely to be overlooked by infrequent testing provided that the emergent behaviour persists over time. In the latter case, however, the rapidity of the onset may be underestimated. Until rather recently, most of the known examples of behavioural change with increasing age were of this type, but more recent studies of development of the chick have revealed frequent and rapid transitions in behaviour over the first two weeks posthatching. In fact, study of age-dependent changes in chicks has made it clear that the development of behaviour is not always gradual.

Thermoregulation and changes in behaviour

At hatching the chick is unable to regulate its body temperature, and this has a marked influence on not only its own behaviour but also the parental behaviour of the hen (Sherry, 1981). Young chicks seek warmth to maintain their body temperature, usually by huddling next to the hen. In the laboratory, they will work for a heat reward by pecking at a key which operates an overhead heat-lamp (Rogers *et al.*, 1977).

The hen caters for the thermoregulatory needs of the young chicks by brooding them under or alongside her body. The chicks also rely on contact with each other for insulation from heat loss. Such huddling behaviour is commonly observed in chicks raised in incubators.

The time spent brooding by the hen declines over the first week posthatching, along with the development of thermoregulatory ability by the chicks (Sherry, 1981). In fact, the mean time spent brooding by the hen declines systematically from day 1 posthatching to a baseline level on day 6. In other words, the chicks spend increasingly less time in physical contact with the hen.

Hence, body temperature of the chicks is initially maintained by behavioural means. As their ability to thermoregulate develops, and the colonic (core) temperature increases from 38.4°C on day 1 to 40.6°C on day 10, they are increasingly more able to tolerate hypothermia in order to feed (Myhre, 1978). According to Myhre (1978), who studied bantam chicks, toleration of hypothermia during feeding does not occur until 5–7 days posthatching. Indeed, days 5–7 is a period when the otherwise gradual rise in colonic temperature shows a transient decline. According to Myhre, this is possibly due to the ending of nutrient supplies from the yolk sac and the consequent need to leave physical contact with the hen for longer periods of time in order to feed, coupled with a lag in the development of full thermoregulatory ability. Broom (1968) would put this stage a day earlier, on day 4, when he showed that moving increases in group-reared chicks. However, he found that pecking at the ground did not increase until day 6. Hogan (1971b) also singles out day 5 as the age by which feeding preferences must be established if the chick is to survive. Days 5–7, therefore, may be a period when competition occurs between feeding and staying warm. After this age, there is a steady development of thermoregulation, determined by metabolic changes and plumage growth (O'Connor, 1984), but it does not appear to be operating fully until day 10 posthatching.

Although the above explanation for the drop in body temperature from day 5 to day 7 has some merit, it seems unlikely to be the complete explanation. Well before day 5 posthatching, the young chicks leave the nest to follow the hen and to peck at small objects even though they may not feed (Chapter 3). This, after all, is the reason why filial imprinting is important. Of course, by following the hen the chicks retain a proximity that ensures contact with her body when they become too cold. Feeding would be possible in the same group context with little risk of hypothermia. The drop in body temperature which the chicks experience from day 5 to day 7 must, therefore, result from increased distance of the chicks from the hen for longer periods of time, possibly as they explore the environment, not only to feed. This does in fact occur on day 5 (Workman and Andrew, 1989; see next section).

It seems probable that behaviour of the hen may often terminate brooding. Thus, when she leaves the nest to feed, the chicks also leave the nest to follow her. Alternatively, as Myhre (1978) suggests, the chicks may terminate brooding by leaving the nest themselves. In fact, Sherry (1981) has estimated that some 20% of the bouts of brooding are terminated by the hen and the rest by the chicks. The relative percentages may well vary with the age of the chicks and consequent changing patterns of interaction between the hen and the chicks. The following more comprehensive study of behavioural development in farmyard chicks enhances our understanding of these and other early transitions in behaviour.

Behavioural transitions in chicks reared by the hen

Workman and Andrew (1989) studied the behavioural development in broods of Warren or Hi-Sex strain chicks raised with the hen in a farmyard. Broods of three male and three female chicks were placed under broody Buff Orpington hens. The groups were observed daily for 60 min until the chicks were 15 days old.

As found previously in jungle fowl (Sherry, 1981), the amount of time spent brooding by the hen declined to a baseline level by day 6 after the chicks had hatched. In this study, brooding was maintained at a high level until day 4 and then declined rapidly. According to Broom (1968), between days 4, 5 and 6 the chicks display increased locomotion, groundpecking, twittering and peeping. This reflects their developing independence at the same time as there is a fall in the hen's brooding behaviour.

Tidbitting by the hen also declined over the first week of posthatching life, with a large decrease between days 7 and 8. This suggests that the hen ceases to direct the chicks' attention to feeding and food sources after the first week of life. Significantly, this is the age at which pecking rate in the pebble-floor tasks increases (Rogers, 1986 and p. 139). The end of the first week might, therefore, be a time of transition from a phase in which feeding is largely controlled by social facilitation, or local enhancement directed by the hen, to a phase of independent food searching and learning by each individual chick to classify food versus non-food objects.

Locomotion by the hen was observed to increase suddenly between days 5 and 6 of the chicks' life, coinciding with the increase in locomotion noted by Broom (1968) in chicks raised without a hen. It remained at the same level from then on. The mean distance of each chick from the hen also increased markedly from day 5 to 6 and then continued to rise gradually, apart from a peak on day 10. This increased separation between the hen and the chicks was not always initiated by the hen. On day 10 the chicks suddenly began to initiate moves by running ahead of the hen. The number of observations of this behaviour was high on days 10 and 11 with a significant decline on day 12. Also, between day 9 and day 10 there was a marked increase in the number of occasions when a chick was out of view of the hen. By day 11 this score had declined and stabilized at a lower level.

Thus, day 10 is a critical age at which the chicks initiate movement away from the hen, even to the extent of spending more time out of her sight. This behaviour occurs less frequently both before and after day 10. There is no obvious reason why day 10 should be a pivotal stage for development of this behaviour, but the earlier research with unilateral injection of cycloheximide into the forebrain hemispheres delineated days 10 and 11 as ages at which key developments are taking place in the right hemisphere. By deduction, it may be hypothesized that a temporary dominance of the right hemisphere begins on day 10 and unmasks these particular behaviours. It is known that the right hemisphere processes spatial information. Therefore, it may not be coincidental that the right hemisphere's temporary dominance on days 10 and 11 coincides with the chicks' independent locomotion using spatial information. Also on day 10, there is a temporary increase in perching behaviour (a peak), which may reflect the use of topographical cues.

It appears that day 10 may mark the beginning of a brief stage of development when each chick forms its own topographical map of its living area. If so, days 10 and 11 may represent a sensitive period for spatial learning, much the same as that for filial imprinting, with the optimal age for spatial learning being on day 10. As mentioned in Chapter 3, the left eye (and right hemisphere) of the chick attends to changes in spatial relations of a bead (Andrew, 1983) and also the left eye is used to guide food-searching behaviour using distant visual cues (Andrew, 1988; Rashid and Andrew,

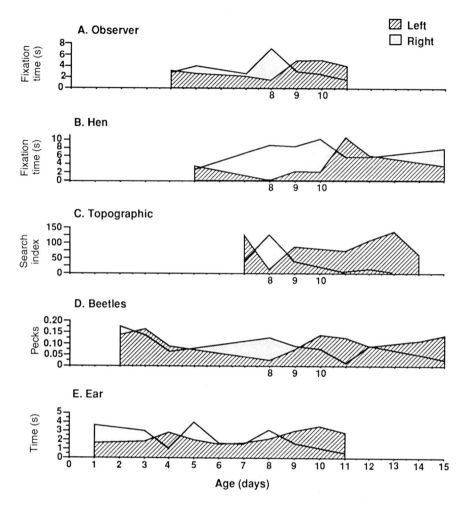

Fig. 5.1. Eye and ear use in chicks of various ages. The hatched area represents the scores for use of the left eye or ear, and the dotted grey area represents the scores for use of the right eye or ear. A, the mean time spent by groups of farmyard chicks using either eye to view a human observer (from Workman and Andrew, 1989); note the preference to use the right eye on day 8. B, similar data for viewing a hen (from Andrew and Dharmaretnam, 1991). C, eye use using topographical cues in searching for food (from Rashid and Andrew, 1989); the search index plotted is a score of successful search using cues either inside or outside the box, a higher score indicating a more successful searching pattern. D, preferred direction of pecking when small beetles are introduced simultaneously into each monocular field (from Andrew and Dharmaretnam, 1991). E, the ear turned toward a sound source emitting a cluck call (from Andrew and Dharmaretnam, 1991). Note the marked transitions of eye and ear use, particularly between days 7 and 12.

1989). From day 9 on, the left eye (right hemisphere) controls such visual searching behaviour (Chapter 3, pp. 106 and 107). All of these results indicate that from around day 10 the left eye and right hemisphere are used for learning about the topographical features of the environment.

Social behaviours may begin to develop earlier, using the right eye and left hemisphere. In fact, they appear suddenly on day 8, exactly when the hen has ceased to perform tidbitting. Workman and Andrew (1989) noticed that suddenly on day 8 the chicks began to stare at, or fixate, a human observer or adult conspecifics other than the familiar hen (Fig. 5.1). Apparently, on day 8 the chick ceases to learn about types of food by watching the tidbitting and feeding of the hen, and instead turns its attention to learning about other aspects of its environment. A similar peak in mean time spent fixating a human observer was found on day 8 in male chicks reared in the laboratory. Moreover, the chicks fixated the observer by using the monocular field of the right eye. This lateralized preference for use of the right eye (left hemisphere) on day 8 is consistent with the use of the right eye by eight-day-old chicks to view a hen (Andrew and Dharmaretnam, 1991; Dharmaretnam and Andrew, 1994; Chapter 3). Presumably, the chicks are using the left hemisphere to categorize the stimulus being viewed (Vallortigara and Andrew, 1991; Vallortigara, 1992b; and pp.87–90). It is noteworthy that the occurrence of this use of the left hemisphere on day 8 coincides exactly with the transient, second peak in cycloheximide susceptibility by the left hemisphere (Fig. 4.1), suggesting that certain neural circuits in the left hemisphere may be undergoing plastic changes in response to stimulation on day 8.

Further social behaviour develops after the behavioural transitions discussed so far. An abrupt increase in frolicking, in which the chick runs with its wings raised or flapping towards another chick, sometimes colliding with it, occurs between days 11 and 12. Kruijt (1964) reported an increase in frolicking and vigorous social interactions on day 10. Sparring does not occur before day 10, but then it increases sharply.

Table 5.1 summarizes the transitions in behaviour discussed so far. It shows clearly that most transitions occur in the period from day 8 to day 10 or 11, and this corresponds to changing susceptibilities to cycloheximide treatment in the left and right hemispheres respectively. Right eye/left hemisphere control appears to dominate on day 8, whereas the left eye/right hemisphere dominates on day 10.

The examples discussed so far apply mainly to male chicks. Females show many of the same transitions in behaviour and at the same ages, but there are some differences. For example, male chicks consistently spend more time out of sight of the hen (Workman and Andrew, 1989). This difference between males and females may be caused by differences in lateralization. Female chicks using the left eye pay less attention to spatial relations than do males using the left eye (Andrew, 1983). Also, age-related changes in the frequency of fear responses given to a violet bead by chicks tested binocularly follow the

Table 5.1 The transitions in behaviour that occur in male chicks at different ages posthatching. Transitions or shifts in behaviour only are included, not continued performance of the same behaviour. Transitions that lead to an increase followed by a decrease in a given behaviour on a particular day are listed as a 'peak' in that behaviour. The changing susceptibility of male chicks to cycloheximide (CXM) is indicated in the right-hand columns for comparison (refer to Fig. 4.1 for details). Here only the transitions in susceptibility to CXM are indicated. The information has been collected from a number of papers, all of which are cited in the text, but particular reference has been made to Workman and Andrew (1989), Workman *et al.* (1991) and Andrew and Dharmaretnam (1991).

Age (days)	Transitions in behaviour	Susceptibility to CXM
1	Standing up develops. Activity increases.	
2	Emerge from under hen for short periods. Visual imprinting may begin, although it could be later. First peak in sleeping. First peak in fear responding to a violet bead in binocularly tested chicks.	Left hemisphere becomes susceptible.
3	Time spent beside hen as well as under her.	
4	Twittering increases.	
5	Active following of hen. Hypothermia occurs while feeding. Peak in sleeping. High levels of fear elicited by binocular viewing of bead. Fear responses greater with right eye viewing. Preference to use right ear to listen to vocalizations.	
6	Moving increases. Start to move further away from hen. Pecking of the ground peaks. Peak in preference for familiar coloured objects related to feeding.	
7	High levels of fear elicited in binocular viewing of bead. Slight peak in sleeping.	CXM has no effects.
8	Faster learning to discriminate food from non-food. Fixation of imprinting stimulus with right eye. Fixation of conspecifics and observer with right eye peaks.	Left hemisphere susceptible again.

Age (days)	Transitions in behaviour	Susceptibility to CXM
8 (cont.)	Temporary control of food searching behaviour by the right eye. Preference for pecking at stimuli in right visual field. Fear responses to violet bead greater with left eye viewing, and binocular viewing elicits low levels of fear.	
9	Shift to left ear preference for listening to vocalizations. Shift to left eye for control of food searching using distant visual cues. Shift away from preference for familiar to novel-coloured objects in feeding.	
10	Thermoregulation fully developed. Peak in running ahead of hen, increased distance from hen and moving out of sight of hen. Perching behaviour peaks. Sparring begins to increase. Marked increase in fear responses to violet bead when viewed binocularly, but not when viewed monocularly. Shift to preference for pecking at stimuli in left visual field.	Right hemisphere susceptible.
11	Large peak in sleeping. Fixation of imprinting stimulus with left eye. Return to preference for familiar-coloured objects related to feeding.	Right hemisphere susceptible.
12	Frolicking behaviour peaks. Fear responses to violet bead have declined.	No further effects of CXM.
13		
14		
15	Perching behaviour increases again. Another possible peak in sleeping.	

same time-course in male chicks using the right eye and female chicks using the left or right eye (Andrew and Brennan, 1984; see next section for more detail). At this point it is simply important to mention that there are sex differences in behavioural transitions apparent as early as day 5 or 6 posthatching (Workman and Andrew, 1989).

Transitions in Fear Behaviour

There are age-dependent changes in the fear responses that individually caged and tested chicks display when presented with a violet bead. Andrew and Brennan (1983) tested male chicks both binocularly and monocularly with a violet bead illuminated by an internal light source. They measured fear responses for 15 sec after presentation of the bead as a cumulative score of shriek and peep calls, jumping at the side of the cage and moving away from the bead. All of these behaviours can be elicited by electrical stimulation of the diencephalon or central mesencephalic grey region (Andrew and Oades, 1973).

On the first time that they saw the bead the chicks tested binocularly showed an early peak in fear responses occurring on day 2, as reported previously (Bateson, 1964b; Kruijt, 1964; Chapter 3). This peak was followed by a trough on day 3 and an increase on day 5 to reach a peak on day 7 (Fig. 5.2). A similar peak in fearfulness on days 7 and 8 has been reported in male chicks tested in the open field (Jones and Black, 1980). On days 8 and 9 the fear responses had declined to another trough, the score being lowest of all ages tested on day 9, only to rise dramatically to a maximum level on day 10.

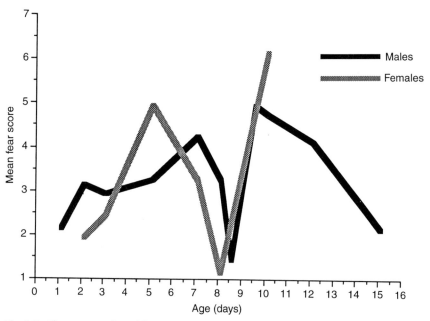

Fig. 5.2. The mean number of fear responses performed by chicks tested binocularly during 15 s following the first presentation of a violet bead. Note the rapid transitions in fear responses at different ages in males (black line) and females (grey line). See text for details. The figure is an adaptation of data from Andrew and Brennan (1983, 1984).

Days 7–9 clearly cover a period of marked transition in fear responding in binocularly tested chicks.

Different patterns of fear responding to the bead were obtained in male chicks tested monocularly at these ages. Those tested using the right eye (left eye occluded) showed a peak in fear responding on day 5, with low levels on day 9 and 10, whereas those tested using the left eye had a trough in fear responding on day 5 and higher levels on days 7 and 8. To summarize, there were significant left eye–right eye differences on day 5, when the level of fear was higher in chicks using the right eye, and on days 7 and 8, when the level of fear was higher in chicks using the left eye (Andrew and Brennan, 1983). Although the fear scores were consistently higher in the (male) chicks tested binocularly compared to monocularly, there was a general tendency for the pattern of change, up until day 9, to follow that of the (male) chicks tested monocularly using the right eye. This was largely confirmed in a later study by the same researchers (Andrew and Brennan, 1984), although it is not consistent with the role of the right archistriatum in controlling fear responses in an unfamiliar environment as found by Phillips and Youngren (1986; and Chapter 4). Of course, this difference in results may depend on the nature of the fear-inducing stimulus or stimuli (bead versus novel environment), as each hemisphere processes different aspects of a visual stimulus (Chapter 3). However, from the findings of Andrew and Brennan (1983, 1984), it may be deduced that over the first nine days posthatching the binocular fear responses of the male chick are controlled by the right eye/left hemisphere. This may explain why there is such a complex pattern of transitions in fear responding over the first two weeks of life. Given that the left and right hemispheres are undergoing sequential developmental changes over this period, these may be reflected in the changing levels of fear responding of binocularly tested chicks, as Andrew and Brennan (1983) suggest.

The pattern for females differs from that of males. Binocularly tested females show no peak fear responses on day 2, but a very marked peak on day 5, with a trough on day 8 and a high level again on day 10 (Fig. 5.2). Monocular testing using either the left or right eye follows essentially the same pattern as that obtained with binocular testing. Compared to males, females have less lateralization for control of fear responses.

The marked changes in fear responding of chicks tested binocularly on days 9 and 10 may indicate changes in hemispheric dominance over this short period. Indeed, the very high level of fear in chicks tested binocularly on day 10 contrasts with low levels in either left- or right-eye monocularly tested chicks (male and female) of the same age. Andrew and Brennan (1983) hypothesized that this may indicate that the hemispheres have become functionally coupled for the first time on day 10. This suggestion is not inconsistent with time-course of development of the supra-optic decussation (Chapter 4).

It may not be insignificant that the sudden increase in fear in the binocular condition on day 10 coincides with increased frolicking and thus more likelihood of the chicks running out of sight of the hen into potentially dangerous, novel situations. Increased fear at this age may ensure that they make their own individual assessments to avoid novel stimuli.

Transitions in Behaviours Related to Feeding

When searching for food, chicks use many different cues, even within the one modality, such as the visual modality. At some ages attention to particular types of information becomes predominant and the pattern of searching for food changes accordingly. Such transitions in searching behaviour are particularly evident at around days 8, 9 and 10 posthatching.

Food searching behaviour using spatial information

In Chapter 3 the use of the left and right eyes in feeding behaviour was discussed. The right eye (left hemisphere) system is used to classify objects into food or non-food; that is, to identify food. The left eye (right hemisphere) system is used to orient to the food source by attending to topographical cues. When both eyes are in use these systems may act together to ensure efficient food searching (Chapter 3). However, it appears that in certain circumstances one system may dominate the other depending on the age of the chick (Rashid and Andrew, 1989).

Chicks were trained binocularly to find a small patch of food buried under sawdust in a tray (Andrew, 1988). They are able to locate the food using two sets of visual cues. Proximal cues were provided by two small bottles placed close to the food and distal cues were provided by the features of the room and the tray itself. After training, the chicks were tested monocularly and with the tray rotated through 180°. Thus, two areas of the tray could be interpreted as indicating the source of food. As mentioned in Chapter 3, from day 9 on the chicks were found to search efficiently at the two areas when using the left eye, and they used both the distal and proximal cues. Chicks of the same age using the right eye were, by contrast, very inefficient in searching. They looked in every corner of the tray.

The same result was obtained in seven-day-old chicks, but day 8 was an exception. On this day the chicks using the right eye searched more efficiently than did those using the left eye, and they did so using proximal cues. This result provides further evidence for day 8 being a special day for development of the left hemisphere, apparently to the extent that it assumes temporary dominance for control of searching behaviour. By deduction, when the chick is tested binocularly, the left eye/right hemisphere system must be dominant at all ages except day 8, when the right eye/left hemisphere system assumes temporary control of responses. Therefore, although no transitions in searching behaviour are greatly apparent in binocularly tested chicks, they are marked in monocularly tested ones (Rashid and Andrew, 1989; Fig. 5.1). Nevertheless, if the searching behaviour is subdivided according to whether the chick is using proximal or distal cues, binocularly tested chicks are found to show a temporary reduction of the use of proximal cues on day 10, when the left eye/right hemisphere with its superior spatial abilities takes over. The reader is reminded that this coincides with the peak in running ahead of the hen and out of her sight presumably under control by the now dominant left eye/right hemisphere.

Transitions in colour preferences

Chicks that have been trained to peck on small coloured boxes to obtain a food reward show changes in their colour preferences with age. Vallortigara *et al.* (1994) trained male and female chicks to peck at the lid of a box so that a drawer containing food would open. The lid of the training box was either green or red. They were tested daily until day 13 with two boxes, red and green, and allowed 20 trials on which their colour choice was determined. There was variation in the preference for the familiar versus the novel colour with age and sex. In both sexes the preference for the familiar colour (red or green) was strongest on day 5 and particularly on day 6, the age at which feeding begins (discussed earlier in this chapter). In females, the preference for the familiar colour was stronger than in males, apart from on day 9 when males trained on the red box showed a stronger preference for that colour than females. In fact, days 9 and 10 were days on which there was a temporary transition to preference for the green box, this shift being stronger in females. In other words, irrespective of the colour on which the chicks had been trained, their preference for the familiar colour was low on days 9 and 10. Preferential responding to a novel-coloured object could have been temporarily enhanced on days 9 and 10 but, as the decreased preference for the familiar was to a level of around 50%, it could be argued that the chicks simply relaxed the criterion for colour and pecked at the objects using other visual cues.

The latter explanation is consistent with evidence for a transient dominance of the right hemisphere around day 9, 10 and 11 because it is the left hemisphere that attends to the colour of objects (Chapter 3). As discussed in the previous section, at this stage of development there is temporary attention to topographical cues using the left eye and right hemisphere. Apparently, this shift in attention is coupled with a loss of attention to colour cues. As colour cues are most pertinent to chicks at other ages, learning about the topographical arrangement of the environment may require suppression of attention to colour by shifting dominance away from the left hemisphere.

Transitions in Eye and Ear Use

The sudden appearance on day 8 of observing conspecifics and humans using the right monocular field has been mentioned already. Andrew and Dharmaretnam (1991) have devised two tasks for measuring the preferred eye to view a stimulus and they have tested chicks at various ages. Sudden transitions in eye preference occurred with changing age.

The first task involved simultaneously presenting two dead *Tenebrio* beetles into each monocular visual field, by impaling them on the two prongs of a fork and advancing the fork from behind the chick's head (Andrew and Dharmaretnam, 1991). The first beetle at which the chick pecked was scored. On days 2, 3 and 4 there was no left or right preference, but on day 8 the beetle on the right was pecked preferentially (Fig. 5.1). On day 9 there was again no preference, but on day 10 there was a shift to preferring the beetle on the left side. This leftward preference persisted throughout the remainder of the testing period, until day 15.

The second task required the chick to put its head through a hole in the cage wall in order to view a stimulus. The position of the head was recorded by videotaping from both overhead and in front of the chick. By scoring only those fixations which used the preferred angles of viewing (34–39° and 61–66° to the beak; Andrew and Dharmaretnam, 1993, and Chapter 3), they were able to obtain clear measures of eye preference. Some of the chicks had been imprinted on a white sphere (a table-tennis ball) suspended in the home cage. When tested with this as the viewing stimulus, on day 4 there was no eye preference, on day 5 there was some tendency to prefer the right eye but it was not quite significant, on day 8 there was a strong preference to use the right eye, on day 9 again there was no preference and on day 11 a strong left-eye preference emerged. When viewing a live hen the right-eye preference was not apparent until day 8 and it persisted until day 11, when the left eye was preferred (Fig. 5.1B). That is, there was a sudden transition between days 10 and 11 from a right to a left eye preference.

To summarize, there is a strong correspondence in eye preference for pecking the beetle and viewing the hen. A strong right eye bias is present on day 8, with a transition to a left eye bias from day 10 or 11 on. Thus, days 8–11 demarcate a period of major transitions in eye use, presumably reflecting changes in hemispheric dominance as already discussed.

A third task investigated ear preferences by measuring the chick's head orientation adopted when listening to the maternal cluck call (Andrew and Dharmaretnam, 1991). The chick was placed in a small pool of light in the middle of an open field. The cluck call was played outside of a circular shielding curtain and the angle of the chick's head position relative to the sound source was determined 15 sec before the chick approached the sound source. Male and female chicks gave somewhat different results up until day 6, but there were similar general trends for both sexes. From day 9 on, both sexes showed a clear preference for turning the left ear toward the sound source, indicating right hemisphere dominance. Before day 4, there was a preference to turn the right ear to the sound source and on day 8 there was also a slight bias towards the right ear (also on day 5 in males only; Fig. 5.1E). Therefore, a transition from the right ear to the left ear occurs between days 8 and 9.

Although the transitions in head turning in this auditory task presumably reflect developmental changes in the lateralization of auditory and not visual processing, there is remarkable coincidence with the transitions in eye use. Possibly both the visual and auditory side preferences are manifestations of general changes in hemispheric dominance with age. If so, it may be hypothesized that a similar pattern of lateralized changes occurs in the other sensory modalities.

Summary of the Rapid Transition Phases of Development

Almost every day of the chick's life over the first two weeks posthatching is marked by transitions in behaviours and hemispheric dominance. These age-dependent shifts in hemispheric dominance may be important determinants of interactions between the hen and the chicks. Although changes in the behaviour of the hen may occur according to a particular time-course following hatching, it is clear that behaviour of the chicks initiates many of these changes. There are striking correspondences in the timing of key transitions in behaviour and of changes in hemispheric dominance across tasks, in chicks raised in groups or isolation, in the farmyard or the laboratory. Apparently, the pattern of transitions is fundamental to the development of brain and behaviour, and presumably to survival. Indeed, left hemisphere and right eye dominance from day 2 to around day 5 or 6 would ensure the chick's full attention to learning to feed (categorization of food versus non-food). This same pattern is strongly expressed again on day 8 just before the transition to right hemisphere and left eye dominance on day 10, when topographical learning about the environment suddenly becomes important for the more independent chick.

Transitions into Adult Behaviour

In the chicken sexual and aggressive behaviour is dependent on the circulating levels of sex hormones. Elevated levels of testosterone in the plasma lead to increases in attack and copulation (Beach, 1961; Andrew, 1966). Systemic administration of testosterone propionate to castrated cocks elevates the levels of both their sexual and aggressive behaviour and also causes them to crow. Administration of testosterone to young chicks

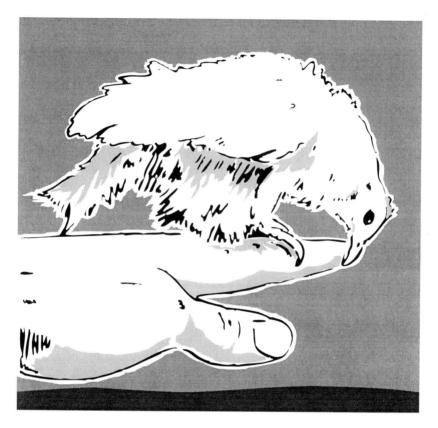

Fig. 5.3. A chick treated with testosterone is displaying juvenile copulation behaviour in the standard hand-thrust test (see text). The chick has mounted the hand, clasped it with the beak, crouched and is performing treading and pelvic thrusting.

increases the incidence of these behaviours precociously, within two or three days after commencement of the treatment with testosterone esters (Andrew, 1966). Andrew (1966, 1975a) has devised methods of testing for attack and copulation in young chicks. These involve using hand-thrust tests. The human hand is used to simulate a hen or another chick. Juvenile copulation involves mounting the hand and clasping it, in the same way that a cock clasps the neck of the hen, followed by pelvic thrusting (Fig 5.3). The horizontal hand, palm facing downwards, is thrust gently at the chick's chest and then held at a level that allows the chick to step on it with ease. The level of copulation is scored according to a rank-ordered scale, ranging from zero for avoiding the hand to 10 for mounting with treading, grasping and pelvic thrusting. Attack behaviour is scored similarly by using the hand as if it were an attacking chick. The palm is held facing the chick with the fingers arched toward the beak. The hand is moved rapidly back and forth near the chick's head, as if it were sparring. The attack score is ranked from zero for averting the gaze to 10 for active sparring with attack leaping and repeated pecks or bites. The chicks are tested in isolation.

After treatment of male or female chicks with testosterone oenanthate, a slow release form of the hormone given intramuscularly (5 or 25 mg given on day 3), Andrew (1975a,b) found elevated copulation, attack, waltzing (a courtship behaviour normally displayed by the adult cock as he circles the hen in a stereotyped manner with the wing nearest the hen lowered; Wood-Gush, 1956) and juvenile crowing. Waltzing and crowing occurred equally in males and females after either dose of testosterone, both rising on about day 7 (Andrew, 1975b). Attack and copulation increased by day 3 following injection in males, but there was no increase in copulation in females (Andrew, 1975a). Thus, even though females are capable of performing the attack and copulation behaviours elicited by hand-thrusting, as evidenced by its occurrence in some, although not all, females (Andrew, 1966), they do not respond to the injected testosterone.

Some of these effects of testosterone are likely to be important to adult hens, as androgens are released by the ovaries. Adult hens do not crow, presumably because their androgen levels are too low, but the cackles produced during laying may be facilitated by androgens (Andrew, 1975b). Circulating androgens also cause reddening of the comb in females as well as males.

Other steroid hormones also elevate the levels of these behaviours. Young and Rogers (1978) found that copulation scores are elevated by oestradiol and 5α-dihydrotestosterone, just as effectively as they are by testosterone, and Balthazart and Hirschberg (1979) found that 5β-dihydrotestosterone acts similarly. Oestradiol treatment of adult capons is also known to restore copulation, although only with partners in the crouched position, not if the partner is uncooperative (Davis and Domm, 1943, cited by Andrew, 1975a). In male Japanese quail (*Coturnix coturnix japonica*) also, oestradiol treatment elevates copulation (Adkins *et al.*, 1980). This result may not be unexpected, since testosterone must be aromatized to oestrogen in the brain (in the hypothalamus in this case) before it can activate the neurons involved in control of copulation (Adkins-Regan, 1983; Hutchison *et al.*, 1986). In general, oestrogenic metabolites of testosterone activate copulatory behaviours and androgenic metabolites active crowing and strutting (Adkins-Regan, 1983). In addition, it appears that oestradiol may have no effect on attack in avian species. Young and Rogers (1978) injected young chicks with oestradiol and found elevated levels of copulation but not attack, whereas in the same experimental paradigm testosterone and 5α-dihydrotestosterone caused increases in both of these behaviours.

During the normal course of development, attack and copulation behaviour in males increases at puberty, when the androgen levels begin to rise. According to the study by Tanabe *et al.* (1979) the plasma testosterone levels of males begin to rise on day 21 posthatching and reach a plateau at

35 days, rising again after day 42 (six weeks). At about six weeks of age, the feral cocks on North West Island, Queensland, Australia, were observed to show agonistic approaches to the maternal hen and to begin attempted mating with the hen (McBride *et al.*, 1969). The attempted mating occurs when the hen is sitting, a posture that allows the young cockerels to mount her. This behaviour was observed to continue for about one week, waning after this time possibly because the hen learns to stand up and move away when a male chick approaches or she meets the chick in an agonistic encounter. Female chicks were not treated in this way. Instead, they were permitted to climb on the hen as they wished.

In male chicks this peak in copulation behaviour early in puberty almost certainly reflects sexual imprinting on the hen (Chapter 3), and it presumably involves learning what may influence sexual behaviour in adulthood. Full adult copulation does not begin to occur until about 120 days after hatching (Kruijt, 1964).

Testosterone clearly has a role in sexual imprinting and it may also have a role in filial imprinting to the hen. Male and female chicks treated with testosterone oenanthate within hours of hatching imprint more strongly on a model hen (Bolhuis *et al.*, 1986). There was a positive correlation between the percentage preference score (Chapter 3) for the hen and the plasma testosterone level. As imprinting on a red box was not similarly affected by testosterone treatment, the testosterone may enhance or accelerate the developing predisposition for the hen (Chapter 3).

Testosterone also alters the attentional processes of chicks. Chicks treated with testosterone are less likely to switch attention from one stimulus type to another, or to be distracted from goal-direct behaviour by extraneous cues. This effect of the hormone was first demonstrated by giving male chicks a choice of red and yellow grains of food scattered among red pebbles, which made their preferred food (red) difficult to find (Andrew and Rogers, 1972). When feeding under these conditions, control chicks switch their attention from red to yellow food at frequent intervals and thus generate frequent short runs of pecking on yellow food grains. Male chicks treated with testosterone show fewer switches to pecking at yellow food grains; they search persistently for red food with long runs of pecking on yellow grains when they do switch attention. In other words, on the rare occasion that a testosterone-treated chick shifts to pecking at yellow food, it continues to search for that colour of food in a persistent manner.

This effect of testosterone on attention is apparent for natural levels of the hormone in adult males (Fig. 5.4). Following castration or treatment with an antiandrogen, cocks peck in shorter runs at different food-types than do control, untreated males (Rogers, 1974). Administration of testosterone to the castrated males causes a return of persistent searching for the red food and, in fact, causes pecking at red grains exclusively (Fig. 5.4). Thus, differences in the circulating levels of testosterone may explain some of the

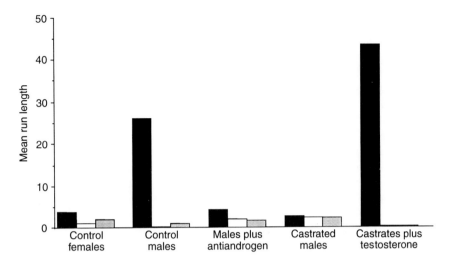

Fig. 5.4. Adult fowl have been tested with a choice of red and yellow food grains scattered on a background of red pebbles. They have been trained to prefer the red food but it is more difficult to find on this background. Control females switch their attention between search for red or yellow food, as shown by their short mean run length of pecking on red grain (black bars) and yellow grain (open bars). Control males search persistently for red food with no pecking at yellow food and only short runs on pebbles (grey bars), presumably pecked while searching for the red food. Note that castration and antiandrogen treatment renders the males less persistent in their searching pattern, and that subsequent administration of testosterone to the castrates reinstates the persistent pattern of search. Data adapted from Rogers (1974).

known sex differences in behaviour of adult fowls.

The attentional persistence, or reduced distractability, of testosterone-treated chicks has also been demonstrated in chicks trained to run down a runway for a food reward (Archer, 1974; Klein and Andrew, 1986). The chicks were trained to run down a runway for a food reward and were then tested with distracting cues (coloured panels) hung on the walls midway along the runway. The controls noticed the panels and stopped to examine them before continuing on down the runway. The testosterone-treated chicks usually failed to stop at the panels, preferring to run on to obtain the food reward. Nevertheless, they apparently noticed the panels, at least in some cases, as some individuals returned to view them after they had eaten for a while (Klein and Andrew, 1986). Changes in the food dish were, by contrast, more disruptive of the performance of testosterone-treated chicks than controls; that is, changes to the goal to which their attention is focused disturbs testosterone-treated chicks to a greater extent than it does controls.

By testing male chicks with moving food dishes, Rogers and Andrew (1989) demonstrated that the testosterone-treated chicks focus their attention on the frontal field. Therefore, they track the moving dishes using short excursions of the head. Controls, by contrast, move their heads through a wide arc as they track using both their monocular and binocular visual fields. Nevertheless, one should not assume that the effect of testosterone is merely to focus attention to the frontal field, as it appears that this may result from the strategy learned by the chick in this particular task. If the dishes containing food are spaced further apart, the testosterone-treated chicks shift their tracking strategy to use large excursions of the head allowing them to peck in the dishes as they move the full breadth of the cage, as previously seen in controls.

In another task requiring chicks to search for grains of food concealed on a checkerboard background, controls used the lateral, monocular field to sight the grain at a distance, whereas the testosterone-treated chicks made much less use of the lateral field and detected the grain at a short distance from its location, using the same strategy they had used in training (Rogers and Andrew, 1989; Fig 5.5). Unlike the controls, the testosterone-treated chicks also displayed persistence by returning to sites where they had previously found food and even to the start, where they had originally been placed in the maze. Apparently, the testosterone-treated chicks specify more stably their plans for search and they are more fixed in their specification of the positions in which food is expected. It is this specification of position that leads testosterone-treated chicks to focus on use of the frontal field in some tasks.

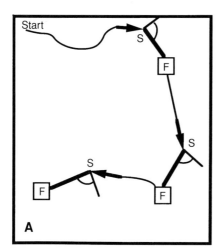
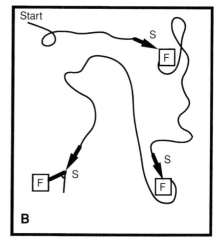

Fig. 5.5. Examples of the searching path adopted by control (A) and testosterone-treated (B) chicks tested on the task with grains of food concealed on a checkerboard pattern. The food is located at three sites (F). The position at which the chick first sighted the food is indicated by S. Controls have a greater sighting distance (S to F). Controls also turn the head to view F using the lateral monocular field, as indicated by the angle at S. Testosterone-treated chicks are more likely to use the frontal field for sighting at a closer distance.

The attentional persistence of testosterone-treated chicks may explain the sex differences in response to testosterone treatment (p. 173). If testosterone fails to produce attentional persistence in females, they may be less likely to focus attention on the hand and therefore be less likely to attack or copulate, as suggested by Andrew (1975a). Similarly, oestradiol treatment may not cause attentional persistence, explaining why its elevation of copulation occurs only when the response can be performed with ease (p. 174). In fact, adult hens switch attention between red and yellow food more frequently than males, which indicates that oestrogen does not cause attentional persistence (Fig. 5.4). Searching strategies used in the natural environment might also be affected by testosterone levels. Males, for example, patrol the boundaries of their territory according to a regular and persistent programme (McBride *et al.*, 1969). It would be interesting to know whether they make frequent returns to certain sites, as did the testosterone-treated chicks in the experimental situation.

The effect of testosterone on attention might also explain why imprinting is stronger in testosterone-treated chicks. More persistent attention to the imprinting stimulus during training and/or testing would be manifested in a higher preference score, but this would not, on its own, explain why the imprinting to the hen and not the red box is influenced by testosterone treatment. It will be recalled that plasma testosterone levels of male chicks rise from a low level on day E17 to reach a peak on day 1 or 2 posthatching (Chapter 2; Fig. 2.6). Tanabe *et al.* (1979) reported a decline in plasma testosterone levels from day 1 to day 3 posthatching. Over the same ages, the plasma testosterone levels of females are lower on days 1 and 2 but rise on day 3 to a level above that of males on day 3 (Tanabe *et al.*, 1979). It would be interesting to know whether these natural fluctuations of hormone levels have consequences on imprinting or any other form of learning. Moreover, potential effects of the other steroid hormones on imprinting have yet to be tested.

Lateralization and hemispheric dominance for attack and copulation

In Chapter 4 it was mentioned that treatment of the left hemisphere of male or female chicks on day 2 posthatching with either cycloheximide or glutamate leads to an elevation of attack and copulation behaviour (Howard *et al.*, 1980; Bullock and Rogers, 1985, 1986). In other words, the unihemispheric drug treatment unmasks these behaviours and reveals lateralization for their control. This result led Rogers *et al.* (1985) to test testosterone-treated chicks monocularly using the standard hand-thrust tests described above. They found that chicks treated intramuscularly with testosterone displayed high levels of copulation when they were tested using the left eye, but when using the right eye they performed as if they had not received an injection of testosterone (Fig. 5.6). This result suggests that neural circuits in the right side of the brain, which receive input from the left eye, activate copulation,

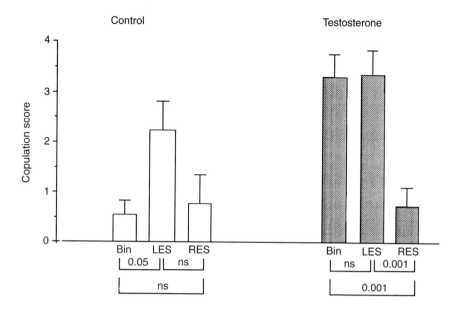

Fig. 5.6. Copulation scores (means with standard errors) are plotted for various groups of male chicks tested on the standard hand-thrust test (Fig. 5.3) from days 6 to 12. The shaded bars represent chicks treated with oestradiol, testosterone or 5α-dihydrotestosterone (all give the same pattern of results). Controls (open bars) were treated with the oil vehicle only. The chicks were tested either binocularly (Bin) or monocularly using the left eye system (LES) or right eye system (RES). Note the elevated scores in treated males tested binocularly and in males using the left eye. The data are adapted from Bullock and Rogers (1985) and Rogers *et al.* (1985).

whereas neural circuits on the left side of the brain suppress this response. These circuits may be in the right hemisphere or right side of the hypothalamus (Barfield, 1969). As the copulation scores of testosterone-treated chicks tested binocularly are equivalent to those of treated chicks tested using the left eye, the treated chicks must have dominance of the right side of the brain (Fig. 5.6). In controls the right eye, and the neural circuits on the left side of the brain that receive input from the right eye, are dominant (Rogers *et al.*, 1985; Fig. 5.6). It is worth noting here that Hutchison *et al.* (1986) have found higher levels of aromatase in the right preoptic region of the hypothalamus in the ring dove, which indicates better conversion of testosterone to its active form, oestrogen, on the right side. Therefore, not only may the right side of the brain assume dominance when testosterone levels are elevated, but subcellular processes on the right side may maximize the action of testosterone in that side.

The same pattern of results is obtained in male chicks treated with oestradiol or 5α-dihydrotestosterone; copulation is elevated when the chicks

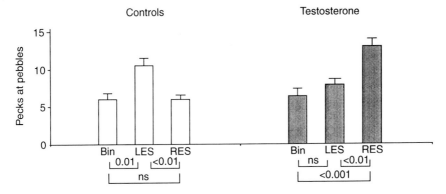

Fig. 5.7. The number of pecks at pebbles in the last 20 pecks of the pebble-floor test in the second week posthatching (Fig. 3.7) has been plotted for control and testosterone-treated (12.5 mg of the oenanthate form on day 2), male chicks. Means and standard errors are plotted for groups of 10 to 12 chicks. The open bars represent the controls, and the shaded bars the testosterone-treated chicks. The chicks were tested binocularly (Bin), using the left eye only (LES) or the right eye only (RES). Note that the testosterone treatment has reversed the eye system (hemispheric) dominance (see text). These data are adapted from Rogers (1986).

are tested using the left eye but not when they use the right eye (Bullock and Rogers, 1992). Moreover, although attack can be tested less easily in monocular chicks, similar results to those for copulation are indicated. Thus, treatment of male chicks with any one of the steroid hormones that elevates attack and copulation appears to reverse hemispheric dominance for control of these behaviours.

The attentional persistence of chicks treated with testosterone might also be a manifestation of right-hemisphere dominance. As mentioned in Chapter 4, treatment of the left hemisphere with cycloheximide causes increased persistence of search for food grains on the pebbled floor, similar to that which follows intramuscular injection of testosterone. Both forms of treatment, by completely different means, may activate the right hemisphere and suppress the left (i.e. reverse the direction of lateralization).

The steroid hormone treatment also reverses hemispheric lateralization for performance on the pebble floor. In control, male chicks, as explained in Chapter 3, performance of chicks using the right eye is superior to that of chicks using the left eye. Testosterone treatment reverses the direction of this lateralization, although the magnitude of the left–right difference is reduced (Zappia and Rogers, 1987; Rogers, 1986). As shown in Fig. 5.7, the number of pecks at pebbles made by binocularly tested, control males in the last 20 pecks of the pebble-floor task (Chapter 3) is equivalent to that of controls tested using the right eye, whereas controls using the left eye make significantly more pecks at pebbles. In fact, the latter do not appear to discriminate grain from pebbles. Following testosterone treatment, binocularly tested chicks have scores equivalent to those of chicks using the left eye, whereas those using the right eye perform more pecking at pebbles. This pattern of results demonstrates dominance of the right eye system (left hemisphere) in controls and dominance of the left eye system (right hemisphere) in testosterone-treated chicks. As found for copulation, testosterone has shifted dominance to the right hemisphere. Indeed, there appears to be a general shift to dominance of the right side of the brain in testosterone-treated chicks, at many levels of neural function. At least, performance on these two apparently distinctly different tasks, copulation and pebble–grain categorization, would indicate a rather general effect of testosterone on the balance of lateralized functions. Behavioural dominance of the left eye (right hemisphere) in testosterone-treated chicks does not necessarily mean that the neural circuits involved in categorizing pebbles and grain have shifted to the right hemisphere. It may simply mean improved interhemispheric access by the right hemisphere to neural circuits located in the left hemisphere. This is a plausible suggestion because interocular transfer of the ability to perform this task occurs readily in either direction (Rogers, 1986). However, why testosterone-treated chicks using the right eye might be unable to access the appropriate neural circuits in the left hemisphere is somewhat paradoxical, unless the testosterone simply suppresses the right eye system and forces interhemispheric access to occur.

By contrast, no interocular transfer occurs for copulation. For example, chicks treated with testosterone show high levels of copulation when tested using the left eye, but when the same chicks are tested using the right eye the performance level declines to the same level as controls. The reverse also occurs; the same chicks that have low levels of copulation when tested using the right eye have high levels when the eye-patch is removed and placed on the other eye (Rogers *et al.*, 1985). Hence, for copulation there is no transfer between the left and right sides of the brain, and the elevation of performance by testosterone occurs through a shift to right-side dominance.

It must be remembered, of course, that these results have been obtained using young chicks treated with pharmacological doses of testosterone. Whether the same effects occur with natural levels of the hormones in older chicks has yet to be determined. Also, the information discussed has been primarily for males. Control female chicks show less lateralization of performance on the pebble-floor task, and testosterone-treatment simply impairs their binocular and monocular performance, irrespective of the eye used (Zappia and Rogers, 1987). Thus, the same hormones affect females and males differently.

Concluding Remarks

Over the first two weeks of posthatching life, independent time-courses of development in each hemisphere appear to generate shifts in hemispheric dominance that are, in turn, correlated with transitions in behaviour. In general, these transition events are more marked in males than females.

Dominance of the left hemisphere for control of behaviour occurs during the first week posthatching with a clear peak of left-hemisphere controlled behaviours on days 5 and 6. At this stage of development, the chick begins to move away from the hen and nest and it attends more earnestly to feeding. Thus, it is not inconsequential that fear responses increase at the same time, as do preferences for familiar colours in feeding responses. Prior to days 5 and 6, the chicks have pecked at grain and learned certain colour preferences, even though this information may not be used for survival until hunger systems have developed, along with the diminution of yolk sac reserves. Also, in preparation for the first need to relinquish constant contact with the hen, the chicks have developed some degree of thermoregulatory ability by days 5 and 6.

Although dominance of the left hemisphere may continue throughout the first week of the chick's life, it may be somewhat relaxed towards the end of that week only to be re-asserted strongly on day 8 just prior to the temporary assumption of dominance by the right hemisphere. On day 8 the left hemisphere is again temporarily susceptible to cycloheximide disruption (see Chapter 4, Fig. 4.1). Its dominance is evident in the preferred use of the right eye (left hemisphere) to observe conspecifics and other complex stimuli, such as humans, that occurs suddenly on day 8.

Day 8 marks the beginning of a period of sudden transitions in behaviour that extends until day 12. Day 10 is the stage of development when a constellation of behaviours change quite suddenly. This appears to depend on a temporary assumption of dominance by the right hemisphere, indicated by preferential use of the left eye and left ear, as well as susceptibility of the right hemisphere to cycloheximide treatment on days 10 and 11. The right hemisphere is used by the chick to construct a topographical map of its environment, at a stage of development when full thermoregulatory ability has been acquired and the chicks begin to run ahead of the hen and even to spend periods of time out of her sight. As on days 5 and 6, these initiatives by the chicks coincide with a marked increased in fear responding, which might well be required as protection during their exploration. Significantly also, attention to colour cues, at least is the context of feeding, is lost during this temporary dominance of the right hemisphere.

The period of right hemisphere dominance around days 10 and 11 is also marked by an increase in social behaviours, including frolicking and sparring. It might be concluded that the right hemisphere plays a greater role than the left in social interaction. Indeed, the experiments with monocular testing of chicks treated with the sex steroid hormones confirm the role of the right hemisphere in activating attack and copulation, which are social behaviours.

Following the administration of pharmacological doses of testosterone, the shift to dominance of the right hemisphere leads to elevated attack and copulation. Therefore, it is possible that the shift to right hemispheric dominance on days 10 and 11 is influenced by changing levels of the sex steroid hormones at that age, but this has yet to be tested. Other learning factors might also be involved.

It should be noted that the increase in attack and copulation performance occurring after treatment with testosterone oenanthate on day 1 or 2 follows a gradual pattern of increase from about day 6 until around days 12 to 14 without any age-dependent transitions in either binocularly or monocularly tested chicks. It seems that, while the hormone level is elevated, the right hemisphere may remain dominant. At least some of the sex differences in the behaviour of adult fowl might, therefore, be associated with differences in hemispheric dominance caused by action of the sex steroid hormones. Presumably, cocks make use of the right hemisphere in agonistic and sexual behaviours and also in making the topographical assessments used in patrolling the boundaries of their territories. Although the right hemisphere might well be used by hens for control of copulatory behaviour and, in fact, they do preferentially use the left eye (right hemisphere) in agonistic encounters (Rogers, 1991), there may be less lateralization in hens (Chapters 2 and 3). Many measures indicate that this may be the case even in young chicks (e.g. for fear responding and performance of the pebble-floor task). It is tempting to suggest that hens may make more balanced, or integrated, use of both the hemispheres in response to their particular environmental and social demands. For example, hens do not patrol territory and they are usually directed to feed by the cock's calling or tidbitting behaviour to potential food objects at a new location. Thus, there may be less demand for the hen to assess topographical information, but more demand to interpret the meaning of calls and to discriminate or categorize food versus non-food using colour and other cues, this being a known function of the left hemisphere.

Of course, the balance of hemispheric control may change with changed hormonal condition in hens (e.g. when the hen becomes broody). Broodiness causes increased agonistic behaviour and changed patterns of feeding (Wood-Gush, 1971), which may possibly be related to changed hemispheric dominance.

Once the chicks have hatched, changing hemispheric dominance and transitions in the behaviour of the hen may influence the behaviour of the chicks. However, it seems that most of the transitions in the behaviour of chicks occur also in chicks raised apart from the hen and even in isolation. Although this strongly implicates the developing chicks themselves as the main initiators of behavioural change, there has so far been insufficient investigation of hen–chick interactions, to be certain of this. Most information on development of the chicks has been obtained in laboratory conditions without the presence of the hen. In Chapter 2 the highly integrated interaction between the hen and the embryos, mediated by vocalizations, was discussed. On the basis of information discussed in the present chapter, one might ask whether the balance shifts toward control by the chicks after hatching.

Even though milestones in the development of the chick may initiate changes in the hen's behaviour, this does not minimize her role in the environment of the chick, ensuring its survival in the natural environment. The programme of the chick's development may determine when it can perform certain kinds of learning but not what it will learn at each particular age.

Finally, it should be recognized that some sudden transitions in behaviour may have consequences on other behaviours. Thus, the transition periods during which the chick actively performs certain forms of learning may either coincide with or be followed by increased rest or sleep. It will be remembered from Chapter 3 that the mean time spent sleeping declines over the first two weeks posthatching, but there are, as yet unexplained, peaks of sleeping particularly on days 2, 5 and 11 (Fig. 3.8). It may be noted that each of these periods of higher amounts of sleeping either follows or coincides with days on which significant learning has taken place: the peak in sleeping on day 2 coincides with or follows filial imprinting; that on day 5 coincides with the beginning of active feeding, although the peak in feeding-related behaviour occurs on day 6, and the one on day 11 follows the period of spatial learning about the environment. At present, one can only focus attention on potentially interesting relationships between changing levels of sleep and transitions in active learning behaviour over the first two weeks posthatching. No conclusions can be reached without more research.

6

COMPARISON WITH DEVELOPMENT IN OTHER SPECIES

Summary

- Altricial species hatch at a more immature stage than precocial species, such as the chick, but they follow the same general programme of development.
- As in the chick, lateralization of brain function is a characteristic of other avian species. For example, the pigeon has lateralization for visual discrimination performance, many parrots have footedness for holding food, and food-storing birds form lateralized memories of the location of their caches.
- A demand for spatial learning and memory appears to increase the size of the hippocampus relative to the rest of the brain. Those species specialized to store food have a larger hippocampus relative to the rest of the brain than do closely related species that do not store food. A larger hippocampus is also found in homing pigeons in comparison to non-homing breeds.
- Many species of songbird are known to have lateralized control of singing. The canary and several other species have left hemisphere dominance for this function, but the zebra finch has right hemisphere dominance.
- In the zebra finch, the left and right hemispheres process different aspects of the species-specific song. The left hemisphere is involved with recognition of the overall pattern of the entire song, whereas the right hemisphere analyses the harmonics in the song syllables.
- In songbirds, the sizes of the song nuclei in the forebrain increase as a result of the addition of new neurons. This occurs when the circulating levels of testosterone rise during the reproductive season. The ability to make new neurons in adulthood is a special characteristic of birds.
- Lateralization of the brain is also characteristic of mammalian species. The presence of the corpus callosum, the interhemispheric commissure connecting the left and right hemispheres, may be important for generating the lateralization of function between the hemispheres. Its development is dependent on the interaction between sex steroid hormones and environmental stimulation, as in the case for the thalamofugal visual neurons in the chick.

Introduction

This chapter addresses the relevance of knowledge about the development of brain and behaviour in the chicken to other species, both avian and mammalian. The aim is to highlight the most important issues. It would be impossible to give a complete discussion of all the differences and similarities between *Gallus gallus* and other avian species. Indeed, avian species have enormous variation in physical structure (Fig. 6.1), habitat and behaviour. Inevitable differences in brain structure and function have evolved with the different environmental demands placed on each species. Like the chicken, some birds have laterally placed eyes with limited binocular overlap in the frontal field and, depending on head size and feather arrangement, some species even have binocular vision behind the head (Fig. 6.2). Others, primarily predators, have frontally placed eyes with a large binocular over-lap (Fig. 6.3) and different forms of conjugate versus non-conjugate eye

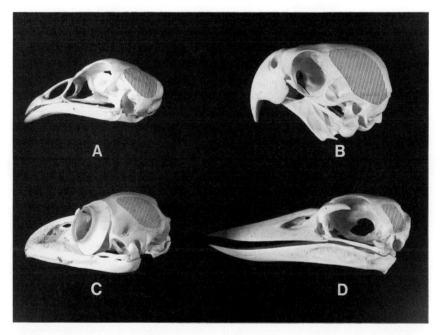

Fig. 6.1. Lateral views of the skulls of the four avian species selected to show the enormous variation in physical structure. Beak structure varies with diet. A, *Gallus gallus*, the chicken, grain feeder. B, *Cacatua galerita*, the sulphur-crested cockatoo, feeding on seeds, nuts, berries, leaf ends and bulbous roots. C, *Podargus strigoides*, the tawny frogmouth, which eats flying insects and small mammals. D, *Dacelo gigas*, the kookaburra, which feeds on snakes, lizards, small mammals and occasionally small birds. The hatched area gives an approximate representation of the brain size; note the variation in size between species. Note also the more frontally directed eye sockets in the predators, particularly in *P. strigoides*.

Fig. 6.2. The palm cockatoo, *Probosciger atterimus*, of northern Australia, Niugini (New Guinea) and Irian Jaya, viewed from the side to the rear of the head to show the large monocular visual field. My observations suggest that it may be possible for this species to have binocular vision behind and above the head, provided that the feathers on the back of the head are not raised and the crest is lowered. Raising of the feathers and crest occurs during social interactions when frontal vision becomes more important. The 360°, or near 360°, vision possible with the crest lowered is used to scan the environment, presumably for predators.

movements (e.g. Wallman and Pettigrew, 1985; Martin, 1985). Along with the large binocular field in predatory birds, there is increased size of the visual Wulst or hyperstriatum, the forebrain region used for binocular vision. The latter is highly developed in owls (Pettigrew and Konishi, 1976a,b; Pettigrew, 1979; Fig. 6.4). Other birds migrate and, therefore, have remarkable abilities to navigate. Navigation requires complex neural analysis and memory processes not present in the chicken. There is no evidence that the hippocampus is used for processing the information needed in navigation, but this region of the forebrain is used in spatial memory formation, as will be discussed in this chapter (Fig. 6.4). The majority of birds have well developed visual and auditory abilities, often impressively so, but some nocturnal species rely more strongly on olfaction than others (e.g. the kiwi, *Apteryx* sp., of New Zealand; Wenzel, 1968). Other species, such as the oilbird, *Steatornis caripensis*, and the Himalayan cave swiftlet, *Collocalia brevirostris*, use ultrasonic echolocation to navigate within the caves where they nest and, in

Fig. 6.3. A blue-footed booby, *Sula nebouxii,* of the Galapagos islands showing frontally directed eyes allowing large binocular overlap. Note also the convergence of the eyeballs to increase this overlap. The same bird has been photographed from two angles to better illustrate this point. This species is predatory upon fish. It has remarkable ability to dive into shallow water and perform fast manœuvres to catch fish.

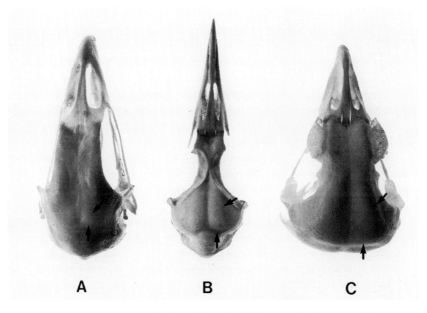

A   B   C

Fig. 6.4. Dorsal views of the skulls of: A, *Gallus gallus* (chicken); B, *Puffinus huttoni* (Hutton's shearwater); C, *Tyto alba* (barn owl). Note the bulging layer of bone over the Wulst or hyperstriatum. The lateral and caudal extent of each right Wulst is indicated by the arrows. Compared to *Gallus, Puffinus* and *Tyto* have enlarged Wulst regions in each case relative to the size of the rest of the brain. Both *Puffinus* and *Tyto* are predators, and *Tyto* is known to have well-developed binocular vision (see text).

addition, the oilbirds appear to use olfactory cues to locate the nest (Snow, 1961). Unlike the chicken, some birds produce complex songs. In these species, there are clearly distinguishable nuclei within the forebrain used for both producing song and assessing the bird's own song and the songs of conspecifics (p. 200). Of all these remarkably varied species, only the chicken, pigeon and owl have been studied by neuroscientists in any degree of detail.

Comparison to Other Avian Species

Precocial versus altricial species

The chick is a precocial species and the pigeon and zebra finch are altricial species. The latter hatch from the egg in a relatively less developed state. Therefore, they must remain in the nest and receive parental care for a longer period of time. Nevertheless, similar forms of early learning may occur in altricial and precocial species. Filial preferences developed by altricial nestlings could be conditioned by food reward, as the parent(s) feed the young in the nest. If so, this would not be imprinting, but Junco (1993) has shown that blackbird nestlings (*Turdus merula*) can develop a preference for a model imprinting object without food reinforcement. This indicates that it is a form of imprinting. Another form of imprinting may occur for song learning in altricial species. Adret (1993) operantly conditioned zebra finches with song as a reward. This procedure strongly influenced song learning during development but not song preferences in adulthood. In adulthood both the conditioned birds and yoked controls, which heard the song but could not interact with it, preferred the training song over any other song. This result suggests the occurrence of a form of imprinting, resulting from exposure to song during a sensitive period of development.

The development of the pigeon and the zebra finch has been studied in some detail. These species are, therefore, the best to compare to the chick. The comparative development of the visual systems will be used as an illustrative example.

The visual systems of the chick are almost fully developed by the time of hatching (Chapters 1 and 2), whereas in the pigeon the same degree of development is not reached until around day 10 posthatching (Fontanesi *et al.*, 1993). Chicks hatch with their eyes open, whereas pigeon squabs do not open their eyes until day 6–9 posthatching, by which age the maturation of the retina is almost complete (Bagnoli *et al.*, 1985). It should be noted that, prior to eye opening, sufficient light passes through the eyelids and nictitating membrane to trigger a pupillary response in pigeon squabs (Heaton and Harth, 1974). During the first week posthatching, the ipsilateral retinotectal projections are lost and the contralateral retinotectal projections enter the superficial layers of the tectum, forming the pattern characteristic of the adult by day 5–6 (Bagnoli *et al.*, 1987; Fontanesi *et al.*, 1993). In chicks, this stage is reached before day E18. Like the pigeon, chicks have ipsilateral retinotectal projections but these are lost before hatching (O'Leary *et al.*, 1983).

The anatomical changes in the development of the pigeon visual system correlate with changes in the distribution of neurotransmitters in the visual neurons, the normal pattern of maturation depending on input from the retina (Fontanesi *et al.*, 1993). As in the chick, the neurotransmitter acetylcholine is present in the pigeon embryo, but other neurotransmitters, including GABA, are not detected until hatching (Bagnoli *et al.*, 1989). Also as in the chick forebrain, the level of GABA in the pigeon hyperstriatum is dependent on visual stimulation. It declines with visual deprivation (Bagnoli *et al.*, 1982).

The zebra finch, *Taeniopygia guttata*, is another altricial species and it hatches at an even more immature stage of development than the pigeon. After an incubation period of only 13 days, it hatches with a very immature visual system that does not respond behaviourally to light stimulation until day 10 posthatching, even though the eyes open by day 5 or 6 posthatching (Bischof and Lassek, 1985). Myelination within the nucleus rotundus, which receives input from the tectum, begins at eye opening and the adult pattern is reached by day 20 to 40 (Herrmann and Bischof, 1986). The size of the nucleus rotundus increases fivefold from hatching to adulthood, and at around day 20 posthatching it undergoes a transient period of being some 20% larger than the adult size (Herrmann and Bischof, 1993). The density of synapses and size of the presynaptic terminals of neurons in the nucleus rotundus increases after hatching, the rate of increase being maximal at eye-opening and for the next few days (Nixdorf and Bischof, 1986). The time course of development of the ectostriatum, to which the nucleus rotundus projects, follows a pattern similar to that of the nucleus rotundus, except that the myelination process is delayed by a few days (Herrmann and Bischof, 1986). As also occurs in the nucleus rotundus, there is a transient period of overproduction of neuronal connections, peaking at around day 20. These developmental events are dependent on visual stimulation (Herrmann and Bischof, 1993).

The over-wiring followed by elimination of neural connections in response to sensory stimulation is characteristic of many developmental processes in birds and mammals alike. In Chapter 4 such a developmental overproduction followed by loss was described for the supra-optic decussation. Again, it will be noted that the overproduction phase occurs earlier in the chick, prior to hatching. The loss of axons in this decussation occurs after hatching, when the chick begins to receive patterned visual stimulation. Overall, therefore, the processes of neural development are similar in precocial and altricial species. It is simply the time courses that differ. However, in the pigeon and other altricial species there may be more overlapping of developmental events occurring in different systems and different brain regions, with less marked transitions in brain development and behaviour compared to development in the chick (cf. Chapter 5). This awaits further, systematic study. To my knowledge, there has been no study of synaptic maturation in the pigeon, and there has been little investigation of the relative rates of maturation of the different sensory systems, as in the chick (Chapters 2 and 3).

Lateralization and memory in other avian species

Until now most research on the development of lateralization has been conducted using chicks, but there is a substantial amount of evidence for lateralized functions being present in other avian species. The evidence for lateralization in pigeons shows them to be remarkably similar to chickens and many species of songbirds have a specialized form of lateralization for control of singing, as will be discussed.

Lateralization in the pigeon (Columba livia)

Adult pigeons show a right-eye advantage in a visual discrimination learning task much like that used for chicks (Güntürkün, 1985; Güntürkün and Kesch, 1987). Pigeons were trained to discriminate grains of safflower seeds from pebbles. The safflower seeds were mixed with pebbles that resembled the grains in their range of colours and shapes. The birds were allowed to feed from a trough containing the mixture and, over several trials, the number of grains ingested was calculated and the number of pecks performed was scored. The pigeons tested using the right eye consumed significantly more seeds than those using the left. They performed more accurately and with a faster pecking rate when they used the right eye (Güntürkün and Kesch, 1987). Like chicks, they were able to categorize grain as different from pebbles better when using the right eye (left hemisphere). A similar result was found when pigeons were trained for pattern discrimination (using a number of patterns) by operant conditioning; the pecking rate was found to be faster for those using the right eye and this was the same in males and females (Güntürkün and Hoferichter, 1985; Güntürkün and Kischkel, 1992).

Lateralization of performance revealed by monocular testing may reflect hemispheric differences in cognitive function or more peripheral lateralization in visual perception. To test between these two possibilities Güntürkün and Hahmann (1994) tested chicks monocularly in a visual acuity task with high-contrast square-wave gratings. The visual acuity was found to be virtually the same for each eye. Thus, the lateralized performance of pattern discrimination must depend on cognitive differences, almost certainly between the hemispheres.

Like the young chicken, adult pigeons also have right-eye dominance for retention performance of a visual discrimination task (von Fersen and

Güntürkün, 1990). In this task the pigeons were trained to discriminate 100 different visual patterns from 625 similar patterns and then they were tested for retention, either binocularly or monocularly. Retention performance was superior when the right eye (left hemisphere) was used. The retention scores were, however, even better when the pigeons were tested binocularly, which indicates that the two eye systems must act in cooperation. In other words, in the pigeon there is no clear hemispheric dominance, such as occurs in the chick. In the chick, binocular performance on most tasks is no better than that of monocular performance using one or the other eye (see Chapter 3), which means that one hemisphere dominates in binocular performance.

It should be emphasized that this form of lateralization on visual tasks is present in adult pigeons, whereas the lateralization revealed in chicks on the pebble-floor task is present only during the first three weeks posthatching. Male chicks tested at three weeks of age show equal ability to perform the task monocularly irrespective of whether they use the left or right eye (Rogers, 1991). Nevertheless, given the marked transitions in lateralization that occur over the first two weeks of the chick's life (see Chapter 5), it may be rash to consider that performance at three weeks remains stable into adulthood. Tests of adult chickens are required before accurate comparison to the pigeon can be made.

Also as in chickens, structural asymmetry is known to be present in the visual projections of pigeons, but in pigeons it is known to be present in the tectofugal system rather than the thalamofugal system (Güntürkün *et al.*, 1989). This has been discussed in Chapters 2 and 4. It will be recalled that the cells in layer 13 of the tectum are larger on the right side, which may mean that they receive more inputs from neurons projecting from the left tectum. The neurons of the optic tecta also receive connections from neurons of two forebrain tracts, the tractus septomesencephalicus and tractus occipitomesencephalicus. Thus, via these two tracts, inputs from the hyper-striatum and archistriatum feed onto the cells in the optic tectum. The cells on the right side may, therefore, be larger because they receive more input from any one, or all, of these regions. Sectioning the tractus occipitomesen-cephalicus also produces differing effects depending on whether the left or right side is sectioned. Lesions of the tract on the left side lead to severe deficits in pecking responses in both monocularly or binocularly tested pigeons, whereas lesions of the tract on the right side have no effect (Güntürkün and Hoferichter, 1985). As the pigeons with the tract on the left side lesioned showed a temporary lack of responsiveness to somatosensory stimulation of the right side of their bodies, the authors suggested that a form of hemispheric neglect is produced by sectioning the tract on the left side (Güntürkün and Hoferichter, 1985).

In pigeons, the lateralization of the pecking response and the size of the tectal cells is present in both males and females (Güntürkün and Kischkel, 1992). This differs from the lateralization in chicks, which occurs to a lesser degree in females and depends on the circulating levels of the sex hormones (Chapter 2). However, it should be remembered that the lateralization that has been studied in chicks is in the thalamofugal, not the tectofugal, visual system.

In chickens the asymmetry in the thalamofugal visual projections is determined by lateralized stimulation of the embryo by light (Chapter 2). Similarly, in the pigeon, light stimulation establishes the structural asymmetry in the tectofugal system (Chapter 2). Pigeons incubated and raised in the dark have no asymmetry in the size of the cells in layer 13 of the optic tecta (Güntürkün, 1990).

Experiments in which the retina of one eye of the pigeon is removed at hatching show that light stimulation affects the balance of thalamofugal visual inputs to the hyperstriatum (Fontanesi *et al.*, 1993), as has been demonstrated by less drastic manipulation in the chick. The side of the thalamus deprived of visual inputs reduces its number of projections to the ipsilateral hyperstriatum, whereas the other side of the thalamus receiving visual inputs increases its number of projections to the contralateral hyperstriatum. The authors did not mention what happens to the contralateral projections from the visually deprived side of the thalamus. Nevertheless, artificially created asymmetrical light stimulation seems to produce asymmetry in the thalamofugal visual system of the pigeon. It would now be of much interest to see whether a similar effect results from the natural asymmetry of light stimulation of the embryo.

There is a need to look at the development of both visual systems within the same species, either the chick or pigeon, as well as to make further comparisons between species. Although the embryos of most species of birds turn the head in a manner similar to that of the chick so that the left eye is occluded, there are some species that do not do so. One of these is the Australian brush turkey, *Alectura lathami*, in which the embryo's head is positioned along the embryo's midline without tilting. The eggs of this species are incubated in mounds of litter and rotting organic material, and thus in darkness (Baltin, 1969). It can therefore be predicted that this species has symmetry to the visual projections and of many of the forebrain functions known to be lateralized in other species.

Lateralization of memory storage in food-storing birds

There is much evidence for lateralized learning and memory storage in chicks (Chapters 3 and 4) and now some elegant experiments have demonstrated similar lateralization of spatial memory in the marsh tit, *Parus palustris* (Clayton, 1993; Clayton and Krebs, 1993).

Marsh tits store food and rely on a well-developed spatial memory to retrieve their stored caches either hours or days later (Sherry, 1989; Krebs, 1990). In the laboratory (Clayton, 1993), marsh tits captured from the wild were deprived of food for 3 h and then released into a room containing a bowl of seeds and artificial trees with perches and small holes for storage. Each hole was large enough to contain only one seed and covered by a string knot to prevent the bird from seeing the seed. The birds were allowed to store five seeds and then they were removed from the room for various intervals of time before being released into it again to retrieve the stored food. In a testing session the number of seeds retrieved in 10 min was scored. First, the tits were trained monocularly using either the right or left eye and then, 3 h after training, they were tested either binocularly or monocularly, using either the same eye or the opposite eye. When the same eye was used in training and in testing, high retrieval scores were obtained, the same as found in binocularly tested birds. When opposite eyes were used in testing and training, the retrieval scores were very low, indicating a lack of interocular transfer after this 3 h retention interval. Simply wearing the eye-patch has no significant affect on the bird's performance, as high retention scores were obtained when the birds were using either the left or right eye at 3 h after binocular training.

The lateralization emerged at a retention interval of 51 h after binocular training. At this time, birds using the right eye in retrieval had high retention scores, but those using the left eye had low scores. Thus, the right eye has better access to the memory, which is apparently located in the left hemisphere at this time. The left hemisphere is apparently the only one associated with long-term storage of the memory. Birds trained and tested using the left eye (right hemisphere) had little or no retention after seven or more hours. The memory had apparently disappeared from the right hemisphere, or it could no longer be accessed from that hemisphere by a neural system connected to the left eye.

Other experiments in which the marsh tits were trained and tested monocularly showed that interocular transfer of the memory was poor at a retention interval of 3 h. At retention intervals of 24 and 51 h, however, there was transfer of memory in birds trained using the left eye and tested using the right eye, but not vice versa (Clayton, 1993). These results provide clear evidence for lateralization of memory storage and transfer in adult food-storing marsh tits. Similar results were obtained by testing the marsh tits on a one-trial associative learning task, which utilized their spatial memory to find hidden pieces of peanut, but did not require them to store the nut first (Clayton, 1993). Thus, the act of storing itself is not essential for forming the memory of the location of the food.

Clayton and Krebs (1993) compared the performance of marsh tits and blue tits, *Parus caeruleus*, on the same one-trial associative learning task. Although closely related to marsh tits, blue tits do not store food. Both species were able to access the long-term memory of the spatial location of food by using neural circuits connected to the right eye (memory in the left hemisphere), as already discussed. However, interocular transfer of memory was found to occur only in the marsh tit. This suggests that storing and non-storing species utilize different mechanisms of memory processing and different forms of access via the two eye systems.

Clayton and Krebs (1993) suggest that interocular transfer of information may be advantageous for a food-storing species because it would allow both eyes to remember visual images, thereby enhancing the richness of the memory stored. This might aid the discrimination of localities needed for well-developed spatial performance. The improvement in interocular transfer with increasing retention interval might also allow the bird to distinguish memories on a temporal basis.

Finally, it is interesting to note that the chicken uses the left eye (right hemisphere) for spatial food searching tasks (Chapter 3 and 5). This appears to be lateralization in the opposite direction to that of marsh tits and blue tits. The difference could depend on whether the tits are using proximal or distal cues (Chapter 5). It would now be interesting to study the temporal aspects of spatial memory in chicks, and interocular transfer at different times after training.

Sandi et al. (1993) found in chicks that interocular transfer of memory of the passive avoidance bead task occurs more readily from the right to the left eye than vice versa (Chapter 4). Furthermore, in an operant visual discrimination task, chicks were found to have better interocular transfer from the right to the left eye (Gaston, 1984). This is the reverse of that found by Clayton (1993) in the food-storing marsh tits. Of course, this is likely to depend on the type of information being processed and stored in the tasks used. The marsh tits were using spatial information, whereas the chicks tested on the passive avoidance or the visual discrimination task were not required to use spatial information. It would certainly be interesting to determine the preferred direction of interocular transfer of memory of various types of tasks in different avian species.

The avian hippocampus and spatial memory

The avian hippocampus, located next to the dorsomedial surface of the telencephalon (Fig. 4.3B), has an important role in spatial learning and memory (Sherry et al., 1992). As might therefore be expected, the hippocampus is larger in food-storing birds. Among the passerine birds, those species that store food have larger hippocampal regions, relative to brain and body size, than the non-storing species (Krebs, 1990). Similarly, among corvids, the food-storing species have larger hippocampal regions, relative to body weight, than do the non-storing species (Healy and Krebs, 1992). Moreover, there is a positive correlation between the extent to which species store food and the relative size of the hippocampus. Apparently, the greater the requirement for spatial learning and memory the larger the volume of the hippocampus relative to the rest of the brain. Other behavioural differences between storing and non-storing species might also account for the differences in hippocampal size, but obvious possibilities such as differences in migratory behaviour, social organization, diet and habitat have been eliminated as possible causes of the differences (Sherry *et al.*, 1992). Indeed, Healy and Krebs (1991) found no relationship between hippocampal volume and migratory ability in passerine birds.

It is particularly interesting that migratory ability bears no relationship to hippocampal size, since migration must require the processing of spatial information and spatial memory. Studies of homing pigeons have shed some light on this apparent paradox. Lesioning the hippocampus of experienced homing pigeons produces no impairment of orientation towards home when the birds are released at a distant site, but homing ability when near the loft is impaired (Bingman and Mench, 1990; Bingman *et al.*, 1990). It appears that the lesioned birds are unable to use familiar local cues to locate their home loft. Thus, it would seem that brain regions outside the hippocampus are used for flight orientation and long-distance navigation, and the hippocampus is used for assessing spatial information about the locality around the home loft. The effect of the hippocampal lesions on the latter ability appears to be specific for a particular form of spatial memory processing because lesioning the hippocampus has no effect on performance of visual acuity and size-difference threshold tasks (Bingman and Hodos, 1992).

These results were obtained when the hippocampal regions of experienced homing pigeons were lesioned. By contrast, if the hippocampus region of young birds is lesioned, they do not acquire the homing ability (Bingman and Mench, 1990; Bingman, 1993). It appears that the hippocampus is necessary for learning how to orient homewards, but once the behaviour has been established it is no longer required.

Consistent with the enlarged hippocampal size in food-storing birds, homing pigeons have a larger relative size of the hippocampus compared to non-homing pigeons (Rehkämper *et al.*, 1988). It would appear that the demand to memorize local spatial cues is the important factor contributing to increased hippocampal size, not navigational ability.

The larger hippocampus in a food-storer, such as the European magpie (*Pica pica*), compared to a non-storer, such as the jackdaw (*Corvus manedula*) results from the addition of more neurons during the rapid growth phase from the nestling stage to adulthood, followed by a lesser loss of neurons with increasing age (Healy and Krebs, 1993). There is no difference between these species in the density of neurons in the hippocampus. A similar result has been obtained for storing versus non-storing passerines. Most importantly, being given the opportunity to store and retrieve seeds results in a relatively larger hippocampus (Clayton and Krebs, 1994a, b). Marsh tits allowed to store and retrieve seeds were found to have larger hippocampal regions, with more neurons, than marsh tits prevented from performing this behaviour. The birds given the opportunity to store and retrieve seeds had a stable hippocampal size, whereas those unable to do so lost neurons in the hippocampus and the region consequently decreased in size. Thus, use of the hippocampus for performing the species-specific food-storing behaviour maintains its size.

The function of the hippocampus in the chick (*Gallus gallus*) is not known, largely because it has not been studied. In Chapter 4, the effect of hippocampal lesions on performance of the passive avoidance bead task was mentioned. Sandi *et al.* (1992) showed that bilateral and left-side-only lesions placed in the hippocampus blocked recall of memory of the task at 18 h after training. This effect of lesioning the hippocampus has no obvious relationship to spatial learning, but the lateralization is interesting. It might well be interesting to investigate a possible lateralized role of the hippocampus in food-storing birds or homing pigeons by placing unilateral lesions in the hippocampus. The placement of lesions into the hippocampal region of both hemispheres of young chicks also impairs performance on the pebble–grain categorization task (S.A. McFadden, personal communication). Although successful pecking for grain in this task requires the chick to search in different places, effects of the hippocampal lesions on behaviours not related to spatial ability could also cause this result.

Before leaving this discussion of the avian hippocampus, it is worth noting that in the zebra finch, *Taeniopygia guttata*, Vockel *et al.* (1990) found relatively high concentrations of the enzyme aromatase, which converts testosterone into oestrogen, in the hippocampus and parahippocampal region. This result suggests a possible role of the sex steroid hormones in the functioning and/or the development of the hippocampus. However, apart from one species (Sherry *et al.*, 1993), no sex differences have been found in the size of the hippocampus (Sherry *et al.*, 1992).

The avian hippocampus is considered to be functionally equivalent to the mammalian hippocampus, even though there are differences in the anatomical organization of these two structures (Erichson *et al.*, 1991; Sherry *et al.*, 1992; Bingman, 1993). There is evidence that the mammalian hippocampus is used in spatial behaviour, mainly from studies using rats (e.g. Morris *et al.*, 1982, 1990). Furthermore, in rats there is lateralization in the size of the hippocampal regions in each hemisphere: in females the left hippocampus is larger and in males the right hippocampus is larger (Diamond, 1985). The larger hippocampus on the right side in male rats, perhaps not insignificantly, correlates with that hemisphere's greater involvement in spatial performance (Denenberg, 1984, and see later).

As in birds, some species of rodents that hoard food are known to have a larger hippocampus than closely related species that do not do so (Sherry *et al.*, 1992). Again the cognitive demand for spatial memory has led to the evolution of a larger hippocampus. Also as in birds, the size of the mammalian hippocampus is affected by environmental demands. In male rats handled in infancy and raised in an enriched environment, the hippocampus on the right side is larger than that on the left side. However, there is no such lateralization in males that have not been handled and have been raised in the impoverished environment of standard laboratory cages (Sherman and Galaburda, 1985). There are clearly many parallels between the avian and mammalian hippocampus. Since much is known about the neurophysiological functioning of the mammalian hippocampus, it might now be used as a basis for conducting similar studies of the avian hippocampus (Bingman, 1992).

Lateralization of song production and perception in songbirds

Chickens and pigeons are not songbirds, and the neuroanatomy of their forebrains differs somewhat from that of songbirds. In songbirds there are several well-delineated nuclei within the forebrain that are used for singing (Fig. 6.5). One of these, known as the higher vocal centre (HVC; formerly known as the hyperstriatum pars caudale) is present in both hemispheres, but lesions of the left HVC, not the right, prevent the bird from being able to sing. Evidence for lateralized control of song production was first discovered by sectioning the nerves supplying the syrinx, the peripheral structure used to produce the song (Fig. 6.5). Nottebohm (1970, 1971, 1972) found that sectioning the left hypoglossus nerve of the adult male chaffinch, *Fringilla coelebs*, results in the loss of most of the components of song, subsong, and calling, whereas sectioning the right hypoglossus affects only a few or none of these components.

Avian vocalizations are produced by expelling air past the elastic membranes of the syrinx, and in songbirds the syrinx consists of two parts, one in each bronchus (Nottebohm, 1972). Each is an independent sound source and the musculature of each part is innervated separately by a branch of the left or right hypoglossus, the tracheosyringealis nerve. In the chaffinch, the musculature on the left side of the syrinx is larger than on the right (Nottebohm, 1971, 1977). There has been some debate about whether each part of the syrinx of songbirds operates independently, an issue relevant to the interpretation of the results of sectioning the left or right hypoglossus nerves. Nowicki and Capranica (1986) have shown that, for producing one of the vocalizations of the black-capped chickadee (*Parus atricapillus*), the two parts of the syrinx are coupled in a non-linear fashion. Yet, although there can be interaction between the two parts, evidence points to each part being functionally decoupled for most vocalizations. This means that most components (syllables) of song are produced by the left side of the syrinx alone. Recently, Hartley and Suthers (1990) have conducted experiments to further test this theory. In canaries (*Serinus canarius*), they plugged the bronchus unilaterally and followed this by denervating the ipsilateral syringeal muscles. Male canaries which had the right bronchus plugged, and therefore were singing with the left side of the syrinx, produced almost normal songs. Those with the left bronchus plugged produced degraded song. Sectioning the right

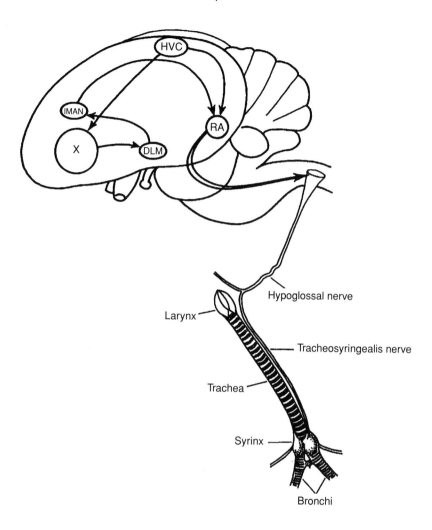

Fig. 6.5. The brain and singing apparatus (syrinx) of a typical songbird, such as the canary. Projections from HVC to RA and from RA to the nucleus of the hypoglossal nerve make up the primary motor output pathway for song production. Regions X, IMAN and DLM are part of recursive loops within song system, important for song learning but not song production (Scharff and Nottebohm, 1991). Auditory input is not included in the diagram; it reaches HVC via the nucleus ovoidalis and Field L (Fig. 4.3). The nuclei are drawn in the left hemisphere only, which is dominant for control of song production in many species (see text), but they are present in both hemispheres. Similarly, only the left hypoglossal nerve and its tracheosyringealis branch, which supplies the left side of the syrinx, is represented. Abbreviations: HVC, higher vocal centre; RA, nucleus robustus of the archistriatum; DLM, medial portion of the dorsolateral thalamic nucleus; IMAN, lateral part of the magnocellular nucleus of the anterior neostriatum. Adapted from Nottebohm (1989).

nerve had little effect on the songs of birds with the right bronchus plugged, but sectioning the left nerve of birds with the left bronchus plugged further impaired song production. Therefore, in canaries, most syllables are sung by the left side of the syrinx alone.

Nottebohm (1977) has shown that in canaries also the left hypoglossal nerve dominates for producing song. As for the chaffinch, sectioning of the left tracheosyringealis nerve impairs singing, whereas sectioning of the right does not. Following sectioning of the right nerve, adult male canaries lost an average of only one-tenth of the song syllables; there were brief silent gaps in the song, noticeable only on sound spectrographs and not directly to the human ear. Sectioning the left tracheosyringealis, however, had a dramatic effect. These birds performed like actors on the silent cinema screen, performing all of the beak, throat, wing, and breathing motor acts usually associated with singing, but producing no sound apart from faint clicks made by uncontrolled vibration of the syringeal membranes as air passed over them. Occasionally, the few remaining syllables of a song issued forth loudly and clearly, to be followed again by breathy silence.

A similar left nerve dominance has also been found in the white-throated sparrow, *Zonotrichia albicollis* (Lemon, 1973), white-crowned sparrow, *Zonotrichia leucophrys* (Nottebohm and Nottebohm, 1976), and the Java sparrow, *Padda oryzivora* (Seller, 1979). Thus, many species of songbird have lateralized control of the syrinx, and dominance of the left nerve and left side of the syrinx.

There are exceptions, however. The zebra finch, *Taeniopygia guttata*, does not have the same direction of lateralization. In zebra finches sectioning the right nerve to the syrinx has a greater effect on song than sectioning the left nerve (Williams *et al.*, 1992).

In some species the lateralization has been traced to the forebrain. In the canary, Nottebohm *et al.* (1976) were able to show that lesioning of the HVC of the neostriatum in the left forebrain hemisphere of the adult canary disrupts song, whereas a similar lesion in the right hemisphere has a much smaller effect (Nottebohm, 1977). It should be mentioned here that the brain centres involved in the control of song largely connect to motor functions on the ipsilateral side, and not the contralateral side, as is more commonly the case for motor functions in vertebrates. There is, as expected, a correspondence between the effects of lesioning the HVC on one side and the effects of sectioning the tracheosyringealis nerve on that same side. Thus, in zebra finches, lesioning the right HVC disrupts song to a greater extent than does lesioning the left HVC (Williams *et al.*, 1992). Almost certainly, therefore, the other four species of songbird known to have dominance of the left nerve must have corresponding dominance of the left HVC.

Following lesioning of the left HVC, canaries are able to produce no more than about one of the syllables of song that they had produced preoperatively. Their song is a monotonous, simple succession of notes with apparently only one syllable remaining. By contrast, following lesioning of the right HVC, they are able to sing one-third to three-fifths or more of the syllables that they had produced preoperatively. Some of the birds with the right HVC lesioned are totally unaffected by the lesion.

Although the HVC is a forebrain nucleus in the motor pathway for control of song (Margoliash and Konishi, 1985), neurons in the HVC display responses to auditory stimulation reaching it from the auditory region of the forebrain, Field L (Kelly and Nottebohm, 1979). In fact, a subset of auditory neurons in the white-crowned sparrow has been shown to be highly selective for the individual's own song (Margoliash and Konishi, 1985). Each bird was reared in the laboratory in a sound-attenuated chamber and exposed to a recorded 'tutor' song. This species learns to sing, but its own song is somewhat different from the tutor's song. It was found that the neurons in the HVC preferred the individual's own song to the tutor's song. As each bird heard the tutor's song prior to singing itself, the specification of these neurons must be delayed until the time when the bird's own song is produced. Since song is learnt by auditory feedback, the bird must listen to its own performance of song and presumably the HVC auditory neurons are involved. That is, HVC neurons must play a role in the development of the motor programme for song. Then, once the individual has learned its song, auditory feedback is no longer required.

Margoliash and Fortune (1992) recorded from HVC neurons in the zebra finch and found units that require two or more syllables of the individual's own song before they will exhibit facilitated responding. Others require combinations of two harmonies with particular frequency and temporal characteristics that were similar to the individual's own song. Thus, it appears that the HVC contains a learned pattern for recognition of the bird's own song. Interestingly, in this study the authors did not state whether they recorded from the left or right HVC, which may be important given the right HVC dominance in the zebra finch.

Neurons in the HVC appear to have both sensory and motor functions. McCasland and Konishi (1981) have also recorded from HVC neurons and they were able to distinguish between the sensory activity of the neurons and the motor activity involved in singing. Comparison of the time courses of these two patterns of neural activity allowed them to show that many spikes of activity, which occurred in response to hearing a song, occurred at times when there were no spikes seen in the motor records. That is, the peaks in auditory activity and motor activity occur at different times. The motor programme controlling the individual's singing appears to be played according to its own temporal sequence and not in response to concurrent sensory feedback. In fact, once the bird has learned its song, the temporal pattern of sensory activity is no longer influenced by auditory feedback, as it occurs even in deafened birds. Thus, once the programme for the individual's song has been established, the neurons display the temporal patterning of the learned song even in the absence of auditory feedback. By having neurons specified for its own song, the bird may be able to recognize the slightly different songs of conspecifics by reference to its own song. Thus, the HVC appears to be the brain site where interaction between sensory and motor activity occurs first for learning and then for maintaining the individual's own song as well as for recognizing conspecific songs.

Degenerating fibres have been traced from HVC lesions to two other nuclei in the forebrain, one called area X in the parolfactory lobe and the other known as the nucleus robustus of the archistriatum (RA) (Nottebohm *et al.*, 1976, Fig. 6.5). In canaries, unilateral lesions of area X are found to have no effect on song, but lesions of the left RA disrupt song (Nottebohm, 1977). The RA has direct connections to the motor nucleus innervating the syrinx (Fig. 6.5). As for the HVC, lesions of the RA on the left side cause greater loss of song than lesions of the RA on the right side. Lesions of either the left or right side lead to a marked reduction in the number of syllables in the song, but lesions of the left RA only result in a reduced frequency range of the fundamental of the song.

Thus, the control of singing in many songbirds is primarily located in specific areas of the left forebrain hemisphere. In canaries there are sex differences in the sizes of area X, HVC, RA and the hypoglossal nucleus. They are all smaller in females. As females do not sing, the size of these nuclei may be related to the singing function. Apart from the hypoglossal nuclei, there is no left–right asymmetry in the sizes of these nuclei. Thus, whereas the hypoglossal asymmetry in structure relates to left hypoglossal dominance for song, higher level asymmetry is functional and not structural. Incidentally, there has been no investigation for potential asymmetries in the organization of the visual system of songbirds, but it would be timely for this to occur, particularly since visual inputs are important to song learning in some species (Chaiken *et al.*, 1993). Also, there has been a report that male zebra finches preferentially use the right eye to view females before copulating (Workman and Andrew, 1986).

There is a period of neural plasticity before one side of the brain assumes dominance for the control of song. In canaries and chaffinches, dominance of the left hypoglossal nerve develops along with the development of stable adult song. Although left hypoglossal section during the first two weeks after hatching causes the right hypoglossus to assume complete control of song, performing this operation during the third and fourth weeks causes incomplete control by the right hypoglossus. Later, when song is developing or has become stable, left hypoglossal section leads to poor singing since the right is much less able to assume control (Nottebohm *et al.*, 1979).

In canaries, but not chaffinches, neural plasticity is reinstated in the next reproductive season when singing resumes (i.e. one year later). In the next season the right side of the brain is able to assume control of singing following sectioning of the left nerve or lesioning of the left HVC. Thus, singing ability

improves. This does not occur in chaffinches because, according to Nottebohm *et al.* (1979), chaffinches, unlike canaries, do not develop new song
repertoires each season. The function of the right HVC in species with
dominance of the left HVC for producing song is not known. Perhaps it is used
in analysing and comprehending the songs of other birds, or in storing a
memory of the individual's own song.

In the zebra finch, which has right HVC and right hypoglossal nerve
dominance for song production (Williams *et al.*, 1992), the HVC regions of
each hemisphere appear to be used to analyse different aspects of the species
song, as indicated following lesioning of the left or right nucleus ovoidalis. The
nuclei ovoidali are situated in the thalamus and they receive auditory input.
Each nucleus ovoidalis projects to Field L in the neostriatal region of its
ipsilateral forebrain hemisphere (Kelly and Nottebohm, 1979), and Field L, in
turn, projects to HVC in the same hemisphere. Thus, removal of the left
nucleus ovoidalis removes auditory input to the left HVC, and vice versa.
These nuclei were lesioned unilaterally in zebra finches and the ability of the
lesioned birds to detect alterations in recorded songs was assessed. Following
lesioning of the right nucleus ovoidalis, the bird took longer to recognize a
missing harmonic in a syllable of a song. Following lesioning of the left
nucleus the bird was less able to discriminate between its own song and that
of a cage mate (Cynx *et al.*, 1992). Therefore, either the left and right HVC
receive a different kind of auditory input or each HVC processes the input
differently.

Nottebohm *et al.* (1990) suggest that, in the zebra finch, the left
hemisphere may be better at discriminating between stimuli which differ in a
variety of ways, and so is involved with 'holistic' perception, whereas the right
hemisphere is involved with the 'analytical' processing of input. It would seem
rather premature to describe the general concepts of holistic versus analytical
to the avian hemispheres on the basis of one piece of evidence, but, of course,
an attempt to draw parallels to the human brain is being made. It should also
be emphasized that control of vocalizations by the zebra finch brain shows
reversed lateralization compared to the several other species of songbirds, and
therefore it may not be representative of birds in general. Moreover,
hypothesized involvement of the left hemisphere in holistic and the right
hemisphere in analytical perception does not appear to fit with the more
extensive data on lateralized perception in the chicken and pigeon. One might
say that the left hemisphere of the chick and pigeon brain appears to perform
analytical processing, whereas the right may be more holistic. The direction
of lateralization may be of very little consequence for a given species on its
own, but, as songbirds evolved after birds without song, such as the chick or
pigeon, one might expect the organization of their hemispheres in some way
to reflect their evolutionary origins. Lateralized involvement of the hemispheres in song production and analysis may have been an elaboration of
lateralization for other functions already present in the avian brain prior to

the evolution of singing ability, or it may have evolved as an independent lateralized system.

To my knowledge no study has been made of hypoglossal control of vocalizations in birds that do not sing, such as the domestic fowl, but there is asymmetry in the organization of the hypoglossal innervation of the syrinx (Youngren *et al.*, 1974). The left hypoglossus innervates the musculature of both the right and left sides of the syrinx, whereas the right hypoglossus innervates only the right side. This organization indicates that there is lateralized control of syringeal function. It would certainly be worthy of investigation.

Birds that mimic would also be interesting to study for potential lateralization. Nottebohm (1976) sectioned the left or right tracheosyringealis nerves in the orange-winged Amazon parrot, *Amazona amazonica*, and found no lateralization. However, unlike songbirds, parrots have branched tracheo-syringealis nerves, each of which supplies both sides of the syrinx. It might therefore be more fruitful to look for lateralization at the forebrain level. Also, it is possible that control of mimicked vocalizations might differ from that for species-specific vocalizations.

Comparison of parrots with other species that mimic, such as mynah birds, might be worthwhile. The HVC nuclei in the forebrain of the mynah bird, *Gracula religiosa*, increase in size over the first year posthatching in parallel with increased ability to mimic (Rausch and Scheich, 1982). Together with these changes there is a decrease in spine density of the neurons and an enlargement of the remaining spines (Rausch and Scheich, 1982). As most of the synapses are on spines, this indicates reduced neuronal inter-connectivity with either increased storage of learned vocalizations or with age. Unfortunately, Rausch and Scheich (1982) made no reference to whether they sampled the left or right HVC regions, or even whether the data were lumped for both sides. Given the potential lateralization, it would be advisable for future researchers to investigate the HVC on the left and right side independently.

Footedness in other avian species

Chickens show a right-foot preference for the first foot movement occurring in a scratching bout during searching for food (Chapter 3). Preferred foot use is even more marked in parrots and cockatoos. Most species prefer to hold food in the left foot while they manipulate it with the beak, and stand on the right foot.

In fact, knowledge of footedness in parrots was recognized three centuries ago (reviewed by Harris, 1989). Friedman and Davis (1938) reported left-footedness for manipulating food in a number of species of African parrots, but the sample size for each species was low. Australian parrots also seem to be predominantly left-footed (Rogers, 1981, 1989). Of nine species scored for foot preference in holding food, eight were significantly left-footed. The remaining species, *Platycercus elegans*, was significantly right-footed. Two other species of the same genus, *Platycercus eximus* and *Platycercus adscitus*, have also been found to be right-footed (Cannon, 1983).

All of the parrots showing footedness feed while perched in trees using the prehensile-footed style. Parrots that do not use the feet in feeding, such as the budgerigar, *Melopsittacus undulatus*, do not show footedness in a task requiring them to remove a small piece of adhesive tape from the beak (Rogers and Workman, 1993). Likewise, pigeons do not use the feet in feeding and they do not show footedness to remove the piece of adhesive tape from the beak (Güntürkün *et al.*, 1988). The footedness that occurs in chickens relates to their use of the feet in feeding, and their right foot preference may be linked to use of the right eye systems to categorize food objects uncovered by scratching. The left- or right-footedness of various species of parrots cannot be similarly explained, but possibly it relates to the type of food eaten by each species (Rogers, 1989).

Although pigeons do not show footedness for removing the adhesive tape, some individual pigeons show significant footedness in landing from flight. Davies and Green (1991) investigated foot use in pigeons during taking off for flight and landing. The pigeons were tested in a flight tunnel and foot use was scored using frame-by-frame analysis of videotapes of these behaviours. No population bias of footedness was found for either taking off or landing, but some individuals had footedness for landing, although not for taking off. The authors proposed that footedness may occur in landing only because landing is a more stressful manœuvre requiring fine visuomotor control. This, they suggested, strongly contrasts to removing a piece of sticking tape from the beak, which uses grooming movements and may be entirely under proprioceptive control. Thus, the individual foot preferences for landing might result from learning a complex manœuvre. This may explain why some individuals are left-footed and others right-footed, although pigeons have lateralization of the visual system and one might therefore have predicted a group bias for visuomotor control. Moreover, at least one other species has been found to show a group bias for footedness for fine, learned manœuvres. Goldfinches, *Carduelis carduelis*, were tested on a task requiring them to manipulate doors and catches using the beak and a foot in order to obtain a food reward. All of the birds preferred to use the right foot (Ducker *et al.*, 1986). These contradictions will not be solved until we know more about the various perceptual, motor and cognitive lateralities within each particular species.

Dependence of brain function and structure on sex hormones

The widespread seasonal variation in agonistic, courtship and reproductive behaviour in avian species demonstrates the importance of the sex steroid hormones in adults. In addition to these seasonal effects, male songbirds are known to have a unique ability to grow new neurons during the reproductive season when singing is required (Nottebohm, 1987). This neuron formation depends on increased levels of testosterone and it occurs in the forebrain nuclei for singing.

It can be said that the songbird brain retains the juvenile ability to form new neurons in adulthood, allowing growth of the song nuclei at times when singing is demanded; that is, the songbird brain retains neural plasticity even in adulthood. Presumably, the growth in certain areas is matched by a loss of neurons in other regions, as occurs in the nuclei for song at the end of the reproductive season. It may be a matter of time-sharing between different brain regions and their functions. This may have evolved to allow the avian brain to remain smaller and lighter to assist flying at the same time as maximizing its functional capacity, as suggested by Nottebohm (1989). If this is correct, avian species that have limited flight (e.g. the chicken) may lack the capacity to form new neurons in adulthood. This hypothesis demands thorough investigation with more comparative information across species. Mammals definitely do not have the capacity for large-scale formation of new neurons in adulthood (Rakic, 1985), although there have been rare sightings of dividing neurons in adult mammals, and recently it has been shown that elevated oestrogen levels induce synaptic plasticity in adult primate neurons even though new neurons do not form (Naftolin *et al.*, 1993).

The remarkable ability to form new neurons in adulthood has been studied mainly in canaries. In male canaries, several of the song nuclei, including the HVC, increase in volume in the spring when new syllables are added to the vocal repertoire. The volume decreases again in the autumn when syllables are lost and the song becomes as unstable as that of juvenile birds (Nottebohm 1989; Nordeen and Nordeen, 1990). Canaries learn a new song repertoire each year (Nottebohm and Nottebohm, 1978) and neurogenesis in the song nuclei is associated with the new repertoire and with the demand for perceiving the song of other canaries during the mating season (Nottebohm *et al.*, 1990).

New neurons that have formed in the HVC grow long axons that project to the RA and thus become a part of the efferent pathway for song control (Kirn *et al.*, 1991). Most of these new projection neurons survive for at least eight months, suggesting that they remain part of the vocal control circuit long enough to participate in the annual renewal of the song repertoire. Thus, the neurogenesis during song development helps to form motor pathways for producing song and it provides synaptic plasticity that may have a role in learning the new song. In other words, the neurogenesis occurs during a sensitive period when new memories of songs are being made. It would be interesting to know more of the role of the new neurons in the processes of memory formation (cf. Chapter 4).

The new neurons form when there are higher circulating levels of testosterone. In fact, autoradiographic studies using zebra finches have shown an accumulation of testosterone in the HVC, the hypoglossal nuclei, and another area connected to the RA, the nucleus intercollicularis (Arnold *et al.*, 1976). Higher concentrations of testosterone-metabolizing enzymes, aromatase and 5α-reductase, are also present in the forebrain nuclei involved in song (Vockel *et al.*, 1990). Thus, uptake and metabolism of testosterone in all the regions involved in singing leads to neurogenesis and a consequent increase in volume of these nuclei. The full song repertoire develops along with this.

The nuclei involved in singing also have receptors for oestrogen (Gahr *et al.*, 1993), which might affect their differentiation. However, as female canaries and zebra finches, for example, do not sing unless they are injected with testosterone, their natural levels of oestrogen do not produce cell growth and singing (Arnold, 1980).

Neurogenesis can also be induced in female canaries and zebra finches by treating them with testosterone (Goldman and Nottebohm, 1983). The size of the vocal repertoire acquired correlates with the volumes of the HVC and RA. The influence of testosterone interacts with specific neural activity within the HVC, as the increase in volume of the HVC in females treated with testosterone also depends on auditory feedback from the individual's own vocalizations. The enlargement of the HVC following testosterone treatment is greatly attenuated in deafened canaries (Bottjer *et al.*, 1986). Thus, in both male and female adult songbirds testosterone promotes neural growth and changes the vocal behaviour pattern. This promotion of neuron growth occurs in interaction with auditory stimulation. That is, the neurogenesis and increased size of the song nuclei responds to stimulus demands, as discussed previously for the hippocampus. Moreover, the interaction between the effects of sex hormones and auditory feedback in causing the increased size of the song nuclei is similar to the interaction between the same hormones and light stimulation in causing developmental changes in the visual pathway of chicks (Chapter 2).

In Chapter 5 an increase in crowing caused by injecting testosterone into young male or female chicks was mentioned. The induction of this call at a neuroplastic stage of development may possibly rely on new neuron growth, as do the vocalizations of songbirds, but this is not known. The seven-day delay between the time of testosterone injection and the increase in crowing (Andrew, 1975b) would be sufficient for new neurons to form and make synaptic connections.

Not only does testosterone influence learning to produce new songs, as discussed so far, but it also influences the bird's ability to discriminate between the bird's own song and the songs of other birds. Cynx and Nottebohm (1992) examined the effects of administering testosterone to castrated zebra finches trained in an operant task to discriminate between songs. The testosterone-treated birds learned to discriminate between their own song and the song of a cage mate in fewer trials than the untreated castrates. As the testosterone had no effect on discriminating between two other zebra finch songs or canary songs, Cynx and Nottebohm (1992) concluded that a raised level of testosterone may help to focus attention on relevant stimuli. Immediately, one is reminded of the now well-documented effect of testosterone in causing attentional persistence in the chick (Andrew, 1991c, Chapter 5). The finding would indicate a need for more cross-fertilization between the chicken and songbird research.

Comparison to Mammalian Species

The thalamofugal visual pathway of birds is considered to be equivalent to the geniculostriate system of mammals and the tectofugal visual system equivalent to the extra-geniculostriate system of mammals (Nauta and Karten, 1970). The visual Wulst or hyperstriatum is equivalent to area 17 of the mammalian visual cortex. Many of the developmental events that are known to occur in the visual systems of birds occur in mammals also. These include the overproduction followed by loss of neural connections, cell death during development and the importance of NMDA receptor expression in neural plasticity (cf. Chapter 4).

The importance of NMDA receptors in the development of mammals is illustrated by the following examples. In the kitten, inhibition of NMDA receptors prevents experience-dependent modification of neural connections in the visual cortex. Kleinschmidt *et al.* (1987) applied the NMDA receptor antagonist APV (aminophosphonovalarate) to the striate cortex of kittens when they had one eyelid sutured. The drug was administered slowly over a period of one week by placing an osmotic minipump into area 17 of the visual cortex. Electrophysiological recordings made at the end of that week revealed that the APV exposed cortex had resisted the effects of monocular deprivation. That is, there was no loss of cells driven by the sutured eye. Inhibition of the NMDA receptors had prevented the neural plastic changes that normally occur in response to monocularity during this sensitive period.

The effects of monocular deprivation in the kitten can be reversed if, within the sensitive period, the originally sutured eyelid is opened and the other eyelid is sutured (Blakemore and van Sluyters, 1974). The cortical neurons shift to being responsive to the newly opened eye. Also, the neurons in the dorsal lateral geniculate receiving input from the deprived eye develop reduced soma sizes compared to those receiving input from the non-deprived eye (Kalil, 1980). This reversal of the effects of monocularity is also dependent on the activity of NMDA receptors, as it can be prevented by APV treatment (Gu *et al.*, 1989).

Similar to the effect of APV, the anaesthetic ketamine–xylazine blocks the cortical cell modifications that occur in the kitten in response to monocularity. Ketamine blocks NMDA receptors, and xylazine is an α-adrenoceptor agonist. Rauschecker and Hahn (1987) gave kittens daily brief (20 min) monocular experience followed by ketamine–xylazine anaesthesia that lasted for 1 h. The treatment, which continued until the kittens had received 30 h of monocular exposure, had a retrograde effect on the cortical plasticity, preventing the shift in ocular dominance. A control group received the treatment after a delay period of 1 h following each monocular exposure, and these kittens showed the expected ocular dominance shift. Subsequent experiments tied the blockage of neural plasticity to the action to ketamine alone without the xylazine (Kossel *et al.*, 1987; Rauschecker *et al.*, 1990). Clearly, as in the chicken, neural plasticity depends on expression of the same neurotransmitter systems. There is considerable evidence that learning and memory formation in mammals likewise depends on NMDA receptor activity, as well as on many of the other neurochemical processes found to be important in the chick (Staubli and Lynch, 1991, and compare with Chapter 4).

Relevance of lateralization in birds to mammalian species

Throughout this book frequent reference has been made to lateralization in the avian brain. The discovery of this lateralization has been extremely important to our understanding of brain mechanisms for control of behaviour and for the study of biological correlates of memory formation. At this stage, however, the reader might well ask: what relevance do all these asymmetries in birds have to other species?

In fact, birds provided the first clear examples of non-human lateralization of the brain. Only later was lateralization confirmed for non-human mammals. Since birds diverged from the line to mammalian (and human) evolution at the level of a common reptilian ancestor, the evidence would point to an early evolution of brain asymmetry. In fact, it is now known that some form of asymmetry existed at the time of the early trilobites. Fossils of trilobites have a higher incidence of predator-inflicted injuries on the right posterior region of their bodies (Babcock and Robison, 1989), indicating that they were either more, or less, successful at escape when they move leftwards, or that their predator had an asymmetry in its direction of attack. Thus, asymmetry of the brain must have evolved very early, soon after the brain structurally duplicated itself. Structural asymmetry is also known in the habenular nuclei of amphibian and lizard brains (Braitenberg and Kemali, 1970; Engbretson *et al.*, 1981), just as it is in the chick (Gurusinghe and Ehrlich, 1985). Furthermore, a recent study has shown that production vocalizations in frogs is controlled by the left hemisphere alone (Bauer, 1993). Thus, it seems that brain lateralization evolved well before the appearance of birds, although avian species may have elaborated on the original theme according to their own particular requirements. It would be interesting to know for what function(s) lateralization evolved originally.

Lateralization in rodents

As we 'ascend' the mammalian branch of the evolutionary tree, we find ample evidence for brain asymmetry in rodents (Denenberg, 1981; Collins, 1985). Although individual rats and mice show right or left paw preferences when they are tested on a task which requires them to reach into a small tube for a food reward, or in their direction of circling (Glick, 1985), there is no overall bias for all individuals to be lateralized in the same direction. Nevertheless, other tests have shown clear lateralization of brain function at a population or group level. The right hemisphere analyses spatial information and controls emotional behaviour (Denenberg *et al.*, 1984). The left hemisphere of mice is known to be used for identifying species-specific vocalizations. The latter is manifested in a right ear (which sends its input to the left hemisphere) advantage shown by the maternal mouse in processing the ultrasonic calls of her pups (Ehret, 1987). Lactating mice were tested in a two-alternative choice task requiring a choice between two sounds played using loudspeakers. One auditory stimulus resembled the ultrasonic distress calls of the pups, and so elicited approach for retrieval of the pups, and the other was a neutral 20 kHz signal. Binocular testing showed a clear choice of the species-specific vocalization, as also did monaural testing with the left ear plugged. Mice with the right ear plugged made no discrimination between the two stimuli. This latter result was not due to an inability to locate the speakers, but rather to an inability to discriminate between the two sounds, because a second experiment showed a choice between two artificial sounds was made equally well with either ear plugged. Therefore, the right ear advantage occurs only for vocalizations important in species-specific communication. This left hemispheric specialization has also been found in Japanese macaques (Petersen *et al.*, 1984), and, of course, for language in humans (Bradshaw and Rogers, 1992, for a review of this literature).

The left hemisphere of the rodent is also used for sequential analysis, whereas the right is used in parallel (one could perhaps call it holistic) analysis. Bianki (1988) has shown that rats using the left hemisphere perform better in conditioning situations when a series of stimuli are presented one after the other, and those using the right hemisphere perform better when the stimuli are presented simultaneously. One notices the similarity to the division of function in the hemispheres of the human brain, but, as discussed previously, it is the opposite to the differential hemispheric processing by zebra finches of their species-specific vocalizations (Nottebohm *et al.*, 1990). Nevertheless, zebra finches have right hemispheric dominance for song control, which is opposite to that of most songbirds.

Lateralization in primates

Contrary to original claims that primates do not have handedness or other forms of lateralization, there is now convincing evidence for a wide range of lateralities in lower and higher primates (Ward and Hopkins, 1993). Reassessment of the data for handedness in primates has revealed significant handedness in many primate species (MacNeilage *et al.*, 1987). Left-handedness is common for visually guided acts of manual prehension and right-handedness for fine manipulation. As in humans, rhesus monkeys express emotions more on the left side of the face (controlled by the right hemisphere) and pay more attention to emotional expressions on the left side of the face of another member of their own species (Sackeim *et al.*, 1978, for humans; Hauser, 1993, for monkeys).

Despite the fact that the visual projections from each eye of the mammal project to both hemispheres, unlike birds in which the optic projections cross over completely, primates show eye preferences. The small-eared bushbaby, *Otolemur garnettii*, has a left-eye preference for viewing a human and food (Rogers *et al.*, 1994b). As in chicks, the preference changes according to the stimulus being viewed or the state of arousal during testing (Chapters 3 and 5). When viewing their babies, the bushbabies shifted their eye preference away from the left eye. The similarity to birds may be due to the fact that in mammals the crossing over projections from the eye to the brain differ from the non-crossing projections, the crossing over fibres being larger and conducting faster that the non-crossing fibres (Bishop *et al.*, 1953). Thus differential input reaches either hemisphere depending on the eye used for viewing. Thus, both birds and higher mammals may show eye preferences that reflect their common evolutionary past. For further discussion of this topic see Ward and Hopkins (1993) and Bradshaw and Rogers (1992).

The corpus callosum and its development

The lateralized functioning of the hemispheres of mammals appears to depend on the corpus callosum, the large collection of axons that connects homologous regions of the left and right hemispheres (Denenberg, 1981). For many cortical functions the left hemisphere may inhibit control by its homologous region in the right hemisphere. Thus, lateralization may be generated by the presence of the corpus callosum. This role differs from that of the tectal and posterior commissures in the chick. As discussed in Chapter 4, the latter commissures in the chick function to suppress lateralization that is revealed only when the commissures are sectioned.

There is some evidence for the corpus callosum being the cause of lateralization at the cortical level but, as yet, it is not conclusive (see Bradshaw and Rogers, 1992, p. 137). Nevertheless, manipulations that affect the degree of lateralization in rats also affect the size of the corpus callosum. For example, handling in neonatal life unmasks some forms of lateralization in behaviour (Denenberg *et al.*, 1981) and it also increases the size of the corpus callosum (Berrebi *et al.*, 1988). Handling involves taking the pups away from their mother for three minutes a day and placing each one separately in a small container. When the pups are returned to the mother she licks them more than usual. This could be the stimulation which affects the development of the corpus callosum, although other aspects of the procedure might also have a role.

The influence of stimulation on the development of the corpus callosum parallels the influence of light stimulation on the development of lateralization in the avian brain (Chapters 2 and 6). In addition, as in the chick, the environmental stimulation interacts with hormonal condition. For example, female rat pups develop an enlarged corpus callosum if they are both handled and injected with testosterone in early life (Denenberg *et al.*, 1991). In females that are allowed to remain with the mother and are not handled, testosterone treatment has no effect on the size of the corpus callosum. Thus, the circulating testosterone influences the size of the corpus callosum only when it interacts with environmental stimulation. The importance of environmental stimulation on development of the corpus callosum is also well recognized in other mammals (Elberger, 1982).

In contrast to the effects of testosterone, the ovarian hormones acting in the neonatal period reduce the size of the corpus callosum (Fitch *et al.*, 1991). Thus, the known sex difference in the size of the corpus callosum in rats (it is smaller in females) appears to be caused by the combined effects of sex hormone levels and early stimulation or experience. Similar processes are likely to occur in higher mammals. Indeed, it may be said that similar interactions between the sex steroid hormone levels and environmental stimulation occur commonly in vertebrates, including birds and mammals, to influence the development of lateralized brain function.

Concluding Remarks

The study of development and function of the avian brain has given rise to several important issues in neuroscience. In each case the findings in birds have been either generally applicable to mammalian species or they have led to advances in knowledge of the mammalian brain by their very difference. On the one hand, the neurochemical events that correlate with memory formation seem to be the same in birds and mammals. Hence, major contributions to this field have been, and still are being made, by studying learning and memory in the young chick (for more detail see Squire *et al.*, 1991; Rose, 1992). On the other hand, it seems that the ability to form large numbers of new neurons in adulthood is unique to avian species, possibly even limited to those species that sing. By understanding the conditions that make new neuron formation possible in the adult avian brain, it may become possible to reinstate neuronal division in the adult mammalian brain, thus providing an exciting potential for repairing brain damage.

The research on hippocampal size in food-storing birds and homing pigeons provides the best illustration of a long-held view that the size of a region in the brain is larger if there has been evolutionary selection for the behaviour controlled by that brain region. Thus, the evolution of food-storing has required superior abilities for spatial learning and memory and therefore there has been genetic selection for a larger hippocampus. Nevertheless, the ontogenetic studies of hippocampal growth show that, within an individual, use of the hippocampus is required if it is to retain its larger size in food-storing birds. In other words, the genetic potential for this region of the brain is expressed only if the environment demands performance of the behaviour that it controls.

A similar interplay between genetic and environmental influences has been revealed by studying lateralization in chick and pigeon brains. In this case, hormonal condition and light stimulation interact with the genetic factors that cause the chick embryo to orient so that it occludes its left eye. As the lateralization that results is due to a complex interaction between all of these factors, it is not possible to isolate one causal factor from another. In rats also, the size of the corpus callosum is the outcome of the interaction between hormonal condition and environmental stimulation in early life.

These two examples illustrate the fallacies in considering only one causal factor for the development of lateralization in the brain. Despite this, researchers of lateralization in the brain persist in attempting to tie the cause of gender differences in humans to the unitary action of the sex hormones in early life (Geschwind and Galaburda, 1987; Kimura, 1992). The studies of factors influencing the development of lateralization in chickens, pigeons and rats provide models for understanding the complex interactions between genes, hormones and environmental factors in determining the final outcome of lateralization in an individual's brain (see Rogers, 1988, for further discussion of these issues).

7

CAN A BRAIN BE DOMESTICATED?

Summary

- Recent research has revealed that birds are capable of complex cognition. In fact, pigeons can perform as well as humans in conditioning tasks using symbol rotation. Most of the evidence available so far is for pigeons and parrots, but there is evidence that chickens interpret the meanings of their species-specific vocalizations and show intention, possibly even deceit, when communicating to members of their own species.
- The long history of domestication of the chicken has led to selection of breeds that are less stressed by being caged or being handled by humans. However, this does not mean that the domestic breeds are well-adapted to living in intensive poultry systems.
- New practices may be implemented to improve welfare and productivity in intensive poultry systems, but it should be realized that even vastly improved intensive systems are unlikely to meet the cognitive demands of the hitherto underestimated chicken brain.

Introduction

It may seem rather inappropriate to discuss issues of animal welfare in a book that has reported many results obtained using invasive procedures, but the information obtained from this research is beginning to change our attitudes to avian species, including the chicken. Although it is recognized that the purpose of much of the research using chicks is couched in terms of its general applicability to vertebrate species (e.g. the investigations of neural plasticity and memory covered in Chapter 4), increased understanding of the chick itself emerges from these investigations. With increased knowledge of the behaviour and cognitive abilities of the chicken has come the realization that the chicken is not an inferior species to be treated merely as a food source.

There has been a tradition of treating birds as cognitively inferior to mammalian species as they have smaller brain to body weight ratios and they lack the neocortex (for more discussion of this point see Eccles, 1992 and Premack, 1978), but recent behavioural research is challenging this concept. The cognitive abilities of some avian species may actually rival those of primates, as will be discussed in the next section. Moreover, the complex cognitive abilities of birds impress us more simply because they can perform these cognitive feats with a smaller brain. Recent findings challenge assumptions that have been made about brain size and the superiority of the mammalian line of evolution.

Cognitive Abilities of Birds

In the primate line of evolution, tool use is considered to be a reflection of higher cognitive function (for more discussion see Gibson and Ingold, 1993). The ability to use tools was once thought to be unique to humans, and linked to having a lateralized brain. Lateralization of the brain was also traditionally thought to be unique to humans (for further discussion see Bradshaw and Rogers, 1992). The latter is now known to be incorrect, as the chapters of this book simply explain. Also, we now know of many examples of tool use by non-human primates and some by birds. For example, some species of finches on the Galapagos Islands use cactus spines to probe into crevices and to impale insects (Millikan and Bowman, 1967). Similarly, great tits have been reported to use pine needles to probe into crevices in bark to retrieve insect larvae (Duyck and Duyck, 1984). Other examples of tool use involving cracking open snail shells by dropping them onto rocks from a height, taking a rock in the beak and using it as a hammer to crack open eggs and, as performed by the green-backed heron (*Butorides striatus*), using bait to attract fish (Sisson, 1974). Recently, Clayton (1994) has observed tool use in the hand-raised marsh tits that she tests on the spatial learning tasks described in Chapter 6. Marsh tits, *Parus palustris*, appear to be compulsive storers because birds fed powdered seed rather than seeds began to remove the stickers from the food containers, to fold them over using the bill and foot and then store them in crevices of their cage. Some of the birds elaborated on this behaviour to include dipping the sticker into the powdered food before storing it. This is tool use according to the definition of Beck (1980).

These examples of tool use by birds indicate that avian species are capable of behaviour thought to be linked to complex cognitive abilities. Chickens cannot be listed among the tool-using birds, but they may display some well-developed cognitive abilities in experimental tests. The finding that chickens can interpret the meaning of their species-specific calls makes an important step towards understanding their cognitive abilities (Evans *et al.*, 1993a; and see Chapter 3). Chickens seem to have intent to communicate, as suggested by the fact that cockerels give alarm calls to aerial predators more frequently when an audience is present (Gyger *et al.*, 1986). In the absence of an audience the cockerels withhold calling, indicating that their response to the aerial predator is not merely reflexive (Chapter 3). Cognitive processes are necessary to make decisions about the social context. Cockerels may even make judgements about their audience and choose to signal dishonestly. There is some evidence that they use the food call in this way to elicit approach by their hen when she has ventured too far away (Marler *et al.*, 1991). More experimentation will be needed to confirm whether this is deliberate deception. If it is, it provides evidence for complex cognitive ability not previously recognized in the chicken. Unfortunately, chicks have not yet been tested in complex cognitive tasks, for example, using operant conditioning and requiring complex decision-making. Therefore, it will be necessary to discuss the research on pigeons that has approached the study of avian cognition in this way. One might venture to predict, however, that chickens might perform similarly to pigeons on at least some of the tasks.

In Chapter 6 it was mentioned that pigeons can acquire object classification skills of hundreds of different stimuli (von Fersen and Güntürkün, 1990). In addition, pigeons have an astounding ability to perform mental rotation problems of the type included in intelligence tests for humans (Delius, 1990; Hollard and Delius, 1982). The pigeons were first trained in an operant conditioning paradigm to perform a matching-to-sample task, by presenting them with stimuli on three keys. The sample (an abstract shape) was presented on the central key, and the two test stimuli on the side keys. One of the test stimuli was identical to the sample and the other was its mirror-image. Pecks at the matching stimulus were rewarded with food, whereas pecks at the mirror-image were punished by a brief period of darkness. Several different shapes were used in the training. In training all of the stimuli were presented at the same angle of orientation. Once the pigeon had acquired this task, it was tested with the comparison shapes rotated at various angles relative to the sample. This rotation task could still be performed just as accurately and as rapidly as the one before. There was no decline in the pigeons' performance when they were asked to include the angular rotation in their assessment. By contrast, humans tested on the same task showed a significant decline in performance, making more errors and requiring longer to react when they were tested with the rotated stimuli. Pigeons are also able to perform as well as humans when tested on tasks requiring discrimination between stimulus arrays on the basis of the number of items contained in each array (Emmerton and Delius, 1993). This was tested in an operant conditioning paradigm with group of dots presented on the keys to be pecked for a food reward. The pigeons had to discriminate one versus two dots and up to seven versus eight dots. The accuracy of discrimination was above 80% for one versus two and declined consistently as the number of dots increased to reach chance levels at seven versus eight. Human subjects tested in the same paradigm also performed at chance levels with groups of seven versus eight dots, suggesting that humans and pigeons process the dot displays in similar ways (Emmerton and Delius, 1993). Pigeons also seem to use the same cognitive processes as do humans for discriminating complex textural differences between surfaces (Cook, R.G., 1993).

Discrimination of textural differences requires perceptual cognition. To make numerosity decisions pigeons may use higher cognitive processes, and for concept formation they may use even higher cognitive processing. Pigeons can perform all of these and they can go beyond this to deal with perceptual concepts, such as trees, leaves, persons, water, fish and even 'sphericity' (Herrnstein, 1982) and they can make mental representations of learned sequences (Terrace, 1985). Their ability to conceptualize 'sphericity' has been determined by conditioning the pigeons using presentations of solid, three-dimensional objects (pebbles, bolts, pearls, buttons, etc.) on a series of metal plates attached to an automated system (Delius, 1985, 1990). The pigeon in the box was presented at any one time with three objects on keys, either two spherical objects and one non-spherical, or one spherical and two non-spherical. They had to peck spherical objects and ignore non-spherical ones. There were 18 objects of each type. After training, the pigeons were tested to see whether they had acquired the concept of 'sphericity' by presenting them with 109 novel spherical and non-spherical objects. They were able to generalize to the novel objects, recognizing them according to the abstract characteristic of 'sphericity', just as do humans. They could even judge sphericity in photographs of the objects. As shown by Watanabe *et al.* (1993) pigeons can recognize objects in coloured photographs with considerable ease.

Performance of these tasks requires the pigeon to have an extensive memory. The pigeons trained by von Fersen and Güntürkün (1990) were able to remember over 600 stimuli of different shapes and they retained that memory with high accuracy over days. More surprisingly, they still retained memory of these stimuli to an 88% accuracy after seven months. Similarly, Vaughan and Greene (1984) have shown that pigeons can remember up to 320 slides of holiday scenes (for humans) for a period of two years. Presumably they achieve this feat by coding or labelling the information, possibly in much the same way that humans do so by using descriptive words (Delius, 1990). Cerella (1986), however, claims that rather simple mechanisms may underlie the enormously complex visual classifications attained by pigeons. Much further experimentation will be necessary to resolve these issues, but these findings have clearly changed our impressions of the cognitive abilities of the pigeon. It would now be interesting to compare these results to other species tested similarly, including the chicken.

The memory capacity of food-storing birds should also be mentioned here. In Chapter 6, the spatial memory ability of tits and corvids was discussed. Some species of tits and corvids in cold regions of the world, such as

Scandinavia and Siberia, store food in the late summer and retrieve it months later during the winter (Shettleworth, 1990). The food is stored in thousands of different locations. As yet, there is no more than circumstantial evidence that these birds have a specific memory of the spatial location of each cache, rather than forming a less specific memory of the general area in which the food has been stored (Shettleworth, 1990). However, field studies with marsh tits, which retrieve their stored seeds only a few hours or days later, have demonstrated that they do form specific memories of the location of each cache. By planting 'control' seeds in areas close to where the tit had stored its seeds, Stevens and Krebs (1986) were able to assess the accuracy of the bird's return to its own cache compared to the control sites. If the tits search at random in a general area, they would be expected to visit and retrieve from control sites with a high frequency. However, the majority of visits were to the bird's own cache, not the control sites. A specific memory of each site seems to be stored, even though the bird spends only a matter of seconds at each site while it is storing the seed. If the Siberian tits form memories of individual sites similarly to these tits, their memory of thousands of locations over months is more than impressive. Perhaps these species have some special adaptations for encoding spatial memory of this kind. Such memories may rely on abstract rules rather than many discrete detailed memories.

There is evidence that pigeons can learn abstract rules, such as that of oddity or difference in terms of the shape of stimuli (Lombardi *et al.*, 1984). The pigeons learn to detect the odd stimulus in a group. The fact that they have learnt the abstract rule is shown by their ability to generalize this rule to other types of stimuli. Earlier studies have shown oddity learning in primates, dolphins and members of the crow family (MacKintosh, 1983), but the original data for pigeons were more equivocal. The recent work by Delius' group has confirmed that pigeons are indeed capable of oddity learning.

Other researchers claim to have shown that pigeons are capable of insight learning equivalent to that of primates (Epstein *et al.*, 1984) and of being aware of their own bodies and own past behaviour (Epstein *et al.*, 1981). Whatever the exact interpretation of these experimental results, it is now clear that birds have cognitive capacities equivalent to those of mammals, even primates. How do they do this with their comparatively small brains? Perhaps it is the special ability of the avian brain to make new neurons in adulthood (Chapter 6).

It is now important to extend these kind of studies to other avian species. The performance of the chicken on these tasks would be a useful comparison, particularly since it would be possible to test very young chicks and so investigate the development of their cognitive abilities. It would be most enlightening to know the ages at which various complex cognitive abilities develop. In addition, as chickens are designed for only limited flight, it would be interesting to compare their problem solving abilities with those of the pigeon, a species for which it would be advantageous to be able to recognize objects at various angles of rotation during flight.

In my opinion, it would also be most interesting to know whether any of these cognitive abilities of the pigeon are lateralized. This information could be obtained simply by using monocular testing, and various combinations of binocularity and monocularity (such as binocular training followed by monocular recall) according to the paradigms used by Clayton and Krebs (Chapter 6).

The important work of Pepperberg (1990a,b) is also changing our attitudes to the cognitive abilities of birds. Pepperberg has shown that parrots (African grey, *Psittacus erithacus*) can be trained to communicate by using words to a competency equivalent to that of the signing chimpanzees. One parrot, Alex, is able to comprehend the concept of symbolic category and respond when asked questions such as 'What colour?' or 'What shape?'. The impressive comprehension skills of Alex are demonstrated by his ability to reply vocally to requests for information about a particular single object in a collection of seven objects (or exemplars). For example, he might be asked 'What colour is the cup?', the cup being one among a collection of differently shaped and coloured objects. Alex responds with approximately an 80% accuracy (Pepperberg, 1990c). Other complex cognitive abilities have also been demonstrated by Alex. These include the ability to respond to questions about similarity and difference based on the relationship between instances of various categories rather than on sets of particular objects (Pepperberg, 1990b), another ability previously thought to be limited to higher primates.

One cannot fail to be impressed by these cognitive capabilities of the parrot. When they are considered in juxtaposition to the known cognitive abilities of pigeons, it seems that such complex cognitive abilities might be rather widespread amongst avian species. At the very least, these recent studies of avian cognition should stimulate active research in the field of comparative cognition. They certainly throw the fallacies of previous assumptions about the inferiority of avian cognition into sharp relief.

Issues of Animal Welfare

I have deliberately chosen to discuss some of the research on cognition in birds before going into issues of animal welfare. To begin with recognition of their cognitive complexity will lead quite logically, I believe, to a demand for improved housing conditions for commercially reared birds. Although many of the currently discussed methods of improving housing for battery hens are most important and definitely to be encouraged, it must be recognized that these are attempts by an industry designed for profit to make some concession to the welfare of the animals. In no way can these living conditions meet the demands of a complex nervous system designed to form a multitude of memories and to make complex decisions. Other researchers have also raised the issue of cognition in discussions of animal welfare. The book by Zayan and Duncan (1987), for example, appealed for combined efforts by experimental psychologists and applied animal ethologists to approach 'the problem of cognition' (page v).

Chickens in battery cages are cramped in overcrowded conditions. Apart from restricted movement, they have few or no opportunities for decision-making and control over their own lives. They have no opportunity to search for food and, if they are fed on powdered food, they have no opportunity to decide at which grains to peck. These are just some examples of the impoverishment of their environment. Others include abnormal levels of sensory or social stimulation caused by excessive tactile contact with cage mates and continuous auditory stimulation produced by the vocalizing of huge flocks housed in the same shed. Also, they have no access to dustbathing or nesting material.

Chickens experiencing such environmental conditions attempt to find ways to cope with them. Their behavioural repertoire becomes directed towards self or cage mates and takes on abnormal patterns, such as feather pecking or other stereotyped behaviours. These behaviours are used as indicators of stress in caged animals (for review see Mason, 1991).

Genetic selection and strain variations

The chicken has a very long history of domestication (Hale, 1969; Appleby *et al.*, 1992) and many have argued that this has led to genetic selection for breeds that are better adapted to commercial farming. This is apparently true. Domesticated breeds have different pituitary functioning from their wild counterparts (Martin, 1978). Domesticated breeds are less fearful, more amenable to accepting new foods and less stressed by closer group contact, but it is a question of degree, not absolutes. I would argue that genetic selection has favoured chickens that can live in farmyard free-ranging conditions in contact with humans and other species, but not in battery cages.

Commercial breeds of chicken may have fewer stress reactions to farmyard conditions than feral or wild strains, but have they lost cognitive complexity? In other words, has the selection been simply for lower brain mechanisms controlling the pituitary adrenal axis rather than higher cognitive functions? This could be tested by comparing commercial and feral strains in operant conditioning procedures, as used in pigeons. Before undertaking such a study it should be remembered that cognitive capacity depends on environmental stimulation throughout development, even on stimulation of the embryo. Therefore, it would not be sufficient to take a battery reared hen and compare it with, say, a jungle fowl raised in more natural conditions. Attempts should be made to equate the rearing conditions as much as possible, although that is not exactly possible because one strain is more reactive or easily stressed than the other.

There are differences in behaviour between breeds of domestic chicks. For example, some breeds are more fearful than others (Jones and Mills, 1983), whereas others vocalize more (Stone *et al.*, 1984) or perch at greater heights (Faure and Jones, 1982). These differences between strains clearly reflect different selective pressures for particular behavioural characteristics, chosen either consciously or inadvertently along with other genetically determined traits, such as feather colour or comb shape. The genetic influences on any particular subset of behavioural characteristics of a breed does not lessen the role of environmental influences. Here one is reminded of the channelling of development by experience, as discussed in Chapter 2. Throughout this book, the interactions between genes and environment have been emphasized. Commercial housing conditions may enhance the development of certain behavioural patterns and suppress others, thus magnifying the apparent differences between domestic breeds and the jungle fowl.

In addition, commercially raised chicks are reared without the hen. Therefore, they are unable to experience the passing on from generation to generation of learned preferences in feeding and possibly other behaviours. Consequently, their cognitive patterns must differ from those of chickens raised with the hen in the natural environment.

Environmental adaptation

It is, of course, possible to take advantage of the behavioural adaptation or learning in order to raise chicks that can deal better with battery caging (i.e. intensive farming). This approach appears to be necessary given the increasing demand for intensively reared chickens to feed the expanding human population. However, it is important not to lose sight of the fact that the suggested improvements in housing and handling practices are minimal concessions for the industry to make.

As mentioned in Chapter 3, Jones (1993) has shown that regular handling of domestic hens, or merely allowing them to see a human for brief periods each day, reduces their fear of humans. This has significance as battery hens are forced to encounter humans at frequent intervals. The outcome of reducing their fear in these ways is increased growth and efficiency of food utilization (Jones and Hughes, 1981), increased resistance to disease (Gross and Siegel, 1982) and reduced aggression (Collins and Siegel, 1987). The birds are more easily captured (Gross and Siegel, 1982) and commercial profits increase with apparently improved welfare of the chickens.

Other abnormal or damaging behaviours that occur in commercially raised fowl, such as feather pecking may be controlled by changing housing conditions. Exactly how to change those conditions to suppress or redirect a particular behaviour pattern requires comprehensive knowledge of the factors that elicit the behaviour. These factors can be assessed most effectively in the wild strain(s) from which the domestic breeds have been derived. For example,

Vestergaard *et al.* (1993) have examined the causes of feather pecking in jungle fowl and found that it is most likely to occur when the birds are dustbathing or, if no dust is available to them, when they show intention movements for dustbathing. More fearful birds were found to have higher levels of feather pecking.

Feather pecking is a major problem in poultry raised in the overcrowded conditions of intensive farming (Blokhuis and Arkes, 1984). The birds peck and pluck out the feathers of their cage mates, particularly the feathers around the anal region. Lower ranking birds suffer most damage from feather pecking. Chiefly, the damage is caused by heat loss and cannibalism (Tauson and Svensson, 1980).

Since increased fear leads to higher levels of feather pecking (Hughes and Duncan, 1972), it can be reduced by lowering the levels of fear. Feather pecking can also be reduced by housing hens on litter rather than wire-flooring (Blokhuis and van der Haar, 1989). Considering all these findings together, it may be concluded that a lack of dustbathing material may lead to more stereotyped intention movements of dustbathing and thus more associated feather pecking. Additionally, a lack of opportunity to dustbathe may raise the level of fear and further contribute to the elevated feather pecking. The practical application of these results would involve a decision not to raise poultry on wire-floors, rather than to use other methods of reducing fear while still keeping them on wire-floor. Therefore, even though chickens may not show particularly high motivation to have access to dustbathing material (Dawkins and Beardsley, 1986), thwarting their performance of this behaviour can affect other behaviours that are damaging. There is much need for detailed analysis of the causal factors involved in other behavioural patterns so that effective solutions can be made to increase welfare and productivity in commercial poultry farming (Dawkins, 1980).

Concluding Remarks

It is becoming increasingly recognized that understanding the cognitive abilities of animals is essential to issues of animal welfare. Yet, there has been insufficient research effort aimed at understanding the cognitive abilities of chickens. In this chapter I was forced to draw on the recent studies of cognition in pigeons in order to make the case for chickens. It is to be hoped that there will be more communication of ideas between scientists interested in cognition and those interested in welfare and production. Exciting developments for agriculture and welfare may emerge from such an exchange. In my opinion, there is a demand to understand the cognitive abilities of the domestic chicken above all avian species, because this bird is the one we have singled out for intensive farming. *Gallus gallus domesticus* is indeed the avian species most exploited and least respected.

REFERENCES

Abe, T. and Matsuda, M. (1992) Developmental change of an enzyme activity oxidizing γ-aminobutyraldehyde to γ-aminobutyric acid in the chick embryonic brain. *Neurochemical Research* 17, 297–299.

Adkins, E.K., Boop, J.J., Koutnik, D.L., Morris, J.B. and Pniewski, E.E. (1980) Further evidence that androgen aromatization is essential for the activation of copulation in male quail. *Physiology and Behavior* 24, 441–446.

Adkins-Regan, E. (1983) Sex steroids and the differentiation and activation of avian reproductive behaviour. In: Balthazart, J., Pröve, E. and Gilles, R. (eds) *Hormones and Behaviour in Higher Vertebrates*. Springer-Verlag, Berlin, pp. 218–228.

Adret, P. (1993) Operant conditioning, song learning and imprinting to taped song in the zebra finch. *Animal Behaviour* 46, 149–159.

Adret, P. and Rogers, L.J. (1989) Sex difference in the visual projections of young chicks: A quantitative study of the thalamofugal pathway. *Brain Research* 478, 59–73.

Adret-Hausberger, M. and Cumming, R.B. (1985) Behavioural aspects of food selection in young chickens. In: Cumming, R.B. (ed.) *Recent Advances in Animal Nutrition in Australia in 1985*. University of New England Press, Armidale, pp. 18.

Adret-Hausberger, M. and Cumming, R.B. (1987a) Social experience and selection of diet in domestic chickens. *Bird Behaviour* 7, 37–43.

Adret-Hausberger, M. and Cumming, R.B. (1987b) Social attraction to older birds by domestic chicks. *Bird Behaviour* 7, 44–46.

Aleksidze, N., Potempska, A., Murphy, S. and Rose, S.P.R. (1981) Passive avoidance training in the young chick affects forebrain α-bungarotoxin and serotonin binding. *Neuroscience Letters* 57, 244.

Ambrosini, M.V., Mariucci, G., Colarieti, L., Bruschelli, G., Carobi, C. and Giuditta, A. (1993) The structure of sleep is related to the learning ability of rats. *European Journal of Neuroscience* 5, 269–275.

Amlaner, C.J. Jr and Ball, N.J. (1988) Avian sleep. In: Kryger, M.N., Roth, T. and Dement, W.C. (eds) *Principles and Practice of Sleep Medicine*. W.B. Saunders, Philadelphia, pp. 50–63.

Andrew, R.J. (1964) Vocalization in chicks, and the concept of 'Stimulus contrast'. *Animal Behaviour* 12, 64–76.

Andrew, R.J. (1966) Precocious adult behaviour in the young chick. *Animal Behaviour* 14, 485–500.

Andrew, R.J. (1975a) Effects of testosterone on the behaviour of the domestic chick. I. Effects present in males and not in females. *Animal Behaviour* 23, 139–155.

Andrew, R.J. (1975b) Effects of testosterone on the behaviour of the domestic chick. II. Effects present in both sexes. *Animal Behaviour* 23, 156–168.

Andrew, R.J. (1976) Attentional processes and animal behaviour. In: Bateson, P.P.G. and Hinde, R.A. (eds) *Growing Points in Ethology*. Cambridge University Press, Cambridge, pp. 95–133.

Andrew, R.J. (1983) Lateralization of emotional and cognitive function in higher vertebrates, with special reference to the domestic chick. In: Ewert, J-P., Capranica, R.R. and Ingle, D. (eds) *Advances in Vertebrate Neuroethology*. Plenum Press, New York, pp. 477–509.

Andrew, R.J. (1988) The development of visual lateralization in the domestic chick. *Behavioural Brain Research* 29, 201–209.

Andrew, R.J. (1991a) The chick in experiment, techniques, and tests. In: Andrew, R.J. (ed.) *Neural and Behavioural Plasticity: The Use of the Domestic Chick as a Model*. Oxford University Press, Oxford, pp. 5–57.

Andrew, R.J. (1991b) The nature of behavioural lateralization in the chick. In: Andrew, R.J. (ed.) *Neural and Behavioural Plasticity: The Use of the Domestic Chick as a Model*. Oxford University Press, Oxford, pp. 536–554.

Andrew, R.J. (1991c) Testosterone, attention and memory. In: Bateson, P.P.G. (ed.) *The Development and Integration of Behaviour*. Cambridge University Press, Cambridge, pp. 171–190.

Andrew, R.J. and Brennan, A. (1983) The lateralization of fear behaviour in the male domestic chick: A developmental study. *Animal Behaviour* 31, 1166–1176.

Andrew, R.J. and Brennan, A. (1984) Sex differences in a lateralization in the domestic chick. *Neuropsychologia* 22(4), 503–509.

Andrew, R.J. and Brennan, A. (1985) Sharply timed and lateralized events at time of establishment of long-term memory. *Physiology and Behaviour* 35, 547–556.

Andrew, R.J. and Dharmaretnam, M. (1991) A timetable of development. In: Andrew, R.J. (ed.) *Neural and Behavioural Plasticity: The Use of the Domestic Chick as a Model*. Oxford University Press, Oxford, pp. 166–173.

Andrew, R.J. and Dharmaretnam, M. (1993) Lateralization and strategies of viewing in the domestic chick. In: Zeigler, H.P. and Bischof, H.-J. (eds) *Vision, Brain, and Behaviour in Birds*. MIT Press, Cambridge, MA, pp. 319–332.

Andrew, R.J. and Oades, R.D. (1973) Escape, hiding and freezing behavior elicited by electrical stimulation of the chick diencephalon. *Brain, Behavior and Evolution* 8, 191–210.

Andrew, R.J. and Rogers, L.J. (1972) Testosterone, search behaviour and persistence. *Nature* 237, 343–346.

Andrew, R.J., Clifton, P.G. and Gibbs, M.E. (1981) Enhancement of effectiveness of learning by testosterone in domestic chicks. *Journal of Comparative and Physiological Psychology* 95, 406–417.

Andrew, R.J., Mench, J. and Rainey, C. (1982) Right–left asymmetry of response to visual stimuli in the domestic chick. In: Ingle, D.J., Goodale, M.A. and Mansfield, R.J.W. (eds) *Analysis of Visual Behavior*. MIT Press, Cambridge, MA, pp. 197–209.

Anokhin, K.V. and Rose, S.P.R. (1991) Learning-induced increase of immediate early gene messenger RNA in the chick forebrain. *European Journal of Neuroscience* 3, 162–167.

Anokhin, K.V., Mileusnic, R., Shamakina, I.Y. and Rose, S.P.R. (1991) Effects of early experience on *c-fos* gene expression in the chick forebrain. *Brain Research* 544, 101–107.

Antal, M. and Polgár, E. (1993) Development of calbindin-D28k immunoreactive neurons in the embryonic chick lumbosacral spinal cord. *European Journal of Neuroscience* 5, 782–794.

Appleby, M.C., Hughes, B.O. and Elson, H.A. (1992) *Poultry Production Systems: Behaviour, Management and Welfare.* CAB International, Wallingford.

Araki, M., Ide, C. and Saito, T. (1982) Ultrastructural localization of acetylcholinesterase activity in the developing chick retina. *Acta Histochemistry and Cytochemistry* 15, 242–255.

Arankowsky-Sandoval, G., Stone, W.S. and Gold, P.E. (1992) Enhancement of REM sleep with auditory stimulation in young and old rats. *Brain Research* 589, 353–357.

Archer, J. (1974) The effects of testosterone on the distractability of chicks by irrelevant and relevant novel stimuli. *Animal Behaviour* 22, 397–404.

Arnold, A., Nottebohm, F. and Pfaff, D.W. (1976) Hormone concentrating cells in vocal control and other areas of the brain of the zebra finch (*Poephila guttata*). *Journal of Comparative Neurology* 165, 487–512.

Arnold, A.P. (1980) Sexual differences in the brain. *American Scientist* 68, 165–173.

Aschoff, J. (1960) Exogenous and endogenous components in circadian rhythms. *Cold Spring Harbor Symposium on Quantitative Biology* 25, 11–28.

Avery, M.L. and Nelms, C.O. (1990) Food avoidance by red-winged blackbirds conditioned with a pyrazine odor. *Auk* 107, 544–549.

Azimi-Zonooz, A. and Litzinger, M.J. (1992) The developing chick brain shows a dramatic increase in the ω-conotoxin binding sites around the hatching period. *International Journal of Developmental Neuroscience* 10, 447–451.

Babcock, L.E. and Robison, R.A. (1989) Preferences of Palaezoic predators. *Nature* 337, 396–696.

Bagnoli, P., Burkhalter, A., Visher, A., Henke, H. and Cuenod, M. (1982) Effects of monocular deprivation in choline acetyltransferase and glutamic acid decarboxylase in the pigeon visual Wulst. *Brain Research* 247, 289–302.

Bagnoli, P., Porciatti, V., Lanfranchi, A. and Bedini, C. (1985) Developing pigeon retina: light-evoked responses and ultrastructure of outer segments and synapses. *Journal of Comparative Neurology* 235, 384–394.

Bagnoli, P., Porciatti, V., Fontanesi, G. and Sebastiani, L. (1987) Morphological and functional changes in the retinotectal system of the pigeon during the early posthatching period. *Journal of Comparative Neurology* 256, 400–411.

Bagnoli, P., Fontanesi, G., Steit, P., Domenici, L. and Alesci, R. (1989) Changing distribution of GABA-like immunoreactivity in pigeon visual areas during the early posthatching period and the effects of retinal removal on tectal GABAergic systems. *Visual Neuroscience* 3, 491–508.

Bakhuis, W.L. and Bour, H.L.M.G. (1980) The behavioural state during climax (hatching) in the domestic fowl (*Gallus domesticus*). *Behaviour* 73, 77–105.

Bakhuis, W.L. and van de Nes, J.C.M. (1979) The causal organization of climax behaviour in the domestic fowl (*Gallus domesticus*). *Behaviour* 70, 185–230.

Ball, N.J., Amlaner, C.J. Jr, Shaffery, J.P. and Opp, M.R. (1988) Asynchronous eye-closure and unihemispheric quiet sleep of birds. In: Koella, W.P., Schulz, H., Obála, F. and Visser, P. (eds) *Sleep '86*. Gustav Fischer Verlag, Stuttgart, p. 127.

Balthazart, J. and Hirschberg, D. (1979) Testosterone metabolism and sexual behaviour in chicks. *Hormones and Behavior* 12, 253–263.

Balthazart, J. and Schoffeniels, E. (1979) Pheromones are involved in the control of sexual behaviour in birds. *Naturwissenschaften* 66, 55–56.

Baltin, S. (1969) Zur Biologie und Ethologie des Talegalla-Huhns (*Alectura lathami* Gray) unter besonderer Berücksichtigung des Verhaltens während der Brutperiode, *Zeitschrift für Tierpsychologie* 6, 524–572.

Bancroft, M. and Bellairs, R. (1977) Placodes of the chick embryo studied by SEM. *Anatomy and Embryology* 151, 97–108.

Banker, H. and Lickliter, R. (1993) Effects of early and delayed visual experience on intersensory development in bobwhite quail chicks. *Developmental Psychobiology* 26, 155–170.

Barber, A.J. and Rose, S.P.R. (1991) Amnesia induced by 2-deoxygalactose in the day-old chick: lateralization of effects in two different one-trial learning tasks. *Behavioral and Neural Biology* 56, 77–88.

Barfield, R.J. (1969) Activation of copulating behaviour of androgen into the preoptic area of the male fowl. *Hormones and Behavior* 1, 37–52.

Bateson, P.P.G. (1964a) Effect of similarity between rearing and testing conditions on chicks' following and avoidance responses. *Journal of Comparative and Physiological Psychology* 57, 100–103.

Bateson, P.P.G. (1964b) Changes in chicks' responses to novel moving objects over the sensitive period for imprinting. *Animal Behaviour* 12, 479–489.

Bateson, P.P.G. (1966) The characteristics and context of imprinting. *Biological Reviews* 41, 177–220.

Bateson, P.P.G. (1972) The formation of social attachments in young birds. *Proceedings of The XVth International Ornithological Congress*. Leiden, E.J. Brill, pp. 303–315.

Bateson, P.P.G. (1974) Atmospheric pressure during incubation and post-hatch behaviour in chicks. *Nature* 248, 805–807.

Bateson, P.P.G. (1976) Rules and reciprocity in behavioural development. In: Bateson, P.P.G. and Hinde, R.A. (eds) *Growing Points in Ethology*. Cambridge University Press, Cambridge, pp. 401–421.

Bateson, P.P.G. (1978) Sexual imprinting and optimal outbreeding. *Nature* 273, 659–660.

Bateson, P.P.G. (1979a) How do sensitive periods arise and what are they for? *Animal Behaviour* 27, 470–486.

Bateson, P.P.G. (1979b) Brief exposure to a novel stimulus during imprinting in chicks and its influence on subsequent preferences. *Animal Learning and Behavior* 7, 259–262.

Bateson, P.P.G. (1980) Optimal outbreeding and the development of sexual preferences in Japanese quail. *Zeitschrift für Tierpsychologie* 54, 231–244.

Bateson, P.P.G. (1982) Preferences for cousins in Japanese quail. *Nature* 295, 236–237.

Bateson, P.P.G. (1983a) Genes, environment and the development of behaviour. In: Halliday, T.R. and Slater, P.J.B. (eds) *Animal Behaviour* Vol. 3 *Genes, Development and Learning*. Blackwell, Oxford, pp. 52–81.

Bateson, P.P.G. (1983b) Optimal outbreeding. In: Bateson, P. (ed.) *Mate Choice.* Cambridge University Press, Cambridge, pp. 257–277.

Bateson, P.P.G. (1987) Imprinting as a process of competitive exclusion. In: Rauschecker, J.P. and Marler, P. (eds) *Imprinting and Cortical Plasticity: Comparative Aspects of Sensitive Periods.* John Wiley, New York, pp. 151–168.

Bateson, P.P.G. (1990) Is imprinting such a special case? *Philosophical Transactions of the Royal Society of London* 329, 125–131.

Bateson, P.P.G. (1991) Making sense of behavioural development in the chick. In: Andrew, R.J. (ed.) *Neural and Behavioural Plasticity: The Use of the Domestic Chick as a Model.* Oxford University Press, Oxford, pp. 113–132.

Bateson, P.P.G. and Jaeckel, J.B. (1976) Chicks' preferences for familiar and novel conspicuous objects after different periods of exposure. *Animal Behaviour* 24, 386–390.

Bateson, P.P.G. and Seaburne-May, G. (1973) Effects of prior exposure to light on chicks' behaviour in the imprinting situation. *Animal Behaviour* 21, 720–725.

Bateson, P.P.G. and Wainwright, A.A.P. (1972) The effects of prior exposure to light on the imprinting process in domestic chicks. *Behaviour* 42, 279–290.

Bateson, P.P.G., Horn, G. and Rose, S.P.R. (1972) Effects of early experience on regional incorporation of precursors into RNA and protein in the chick brain. *Brain Research* 39, 449–465.

Bateson, P.P.G., Rose, S.P.R. and Horn, G. (1973) Imprinting: lasting effects on uracil incorporation into chick brain. *Science* 181, 576–578.

Bateson, P.P.G., Horn, G. and Rose, S.P.R. (1975) Imprinting: correlations between behavior and incorporation of [^{14}C]uracil into chick brain. *Brain Research* 84, 207–220.

Batuecas, A., Cubero, A., Barat, A. and Ramirez, G. (1987) The GABA$_A$ receptor complex in the developing chick optic tectum: ontogeny of [^{3}H]flunitrazepam and [^{3}H]TBPS binding sites. *Neurochemistry International* 11, 425–431.

Bauer, R.H. (1993) Lateralization of neural control for vocalization by the frog (*Rana pipiens*). *Pyschobiology* 21, 243–248.

Beach, F.A. (1961) *Hormones and Behavior.* Cooper, New York, pp. 33–105.

Bear, M.F. and Singer, W. (1986) Modulation of visual cortical plasticity by acetylcholine and noradrenaline. *Nature* 320, 172–176.

Beck, B.B. (1980) *Animal Tool Behavior,* Garland Press, New York.

Bell, G.A. and Rogers, L.J. (1992) Metabolic activity in the hyperstriatum of 2-day-old chicks during optomotor and contrasting visual stimulation. *Behavioural Brain Research* 50, 177–183.

Berrebi, A.S., Fitch, R.H., Ralphe, D.L., Denenberg, J.P., Friedrich, V.L. Jr and Deneberg, V.H. (1988) Corpus callosum: region-specific effects of sex, early experience, and age. *Brain Research* 438, 216–224.

Bertossi, M., Roncali, L., Nico, B., Ribatti, D., Mancini, L., Virgintino, D., Fabiani, G. and Guidazzoli, A. (1993) Perivascular astrocytes and endothelium in the development of the blood–brain barrier in the optic tectum of the chick embryo. *Anatomy and Embryology* 188, 21–29.

Beuving, G., Jones, R.B. and Blokhuis, H. (1989) Adrenocortical and heterophil/ lymphocyte responses to challenge in hens showing short or long tonic immobility reactions. *British Poultry Science* 30, 75–84.

Bianki, V.L. (1988) *The Right and Left Hemispheres: Cerebral Lateralization of Function.*

Monographs in Neuroscience. Vol. 3, Gordon and Breach, New York.

Bingman, V.P. (1992) The importance of comparative studies and ecological validity for understanding hippocampal structure and cognitive function. *Hippocampus* 2, 213–220.

Bingman, V.P. (1993) Vision, cognition, and the avian brain. In: Zeigler, H.P. and Bischof, H-J. (eds) *Vision, Brain and Behavior in Birds.* MIT Press, Cambridge, MA, pp. 391–408.

Bingman, V.P. and Hodos, W. (1992) Visual performance of pigeons following hippocampal lesions. *Behavioural Brain Research* 51, 203–209.

Bingman, V.P. and Mench, J.A. (1990) Homing behavior in hippocampus and parahippocampus lesioned pigeons following short-distance releases. *Behavioural Brain Research* 40, 227–238.

Bingman, V.P., Ioalé, P., Casini, G. and Bagnoli, P. (1990) The avian hippocampus: evidence for a role in development of homing pigeon navigational map. *Behavioural Neuroscience* 104, 906–911.

Bischof, H-J., and Lassek, R. (1985) The gaping reaction and the development of fear in young zebra finches (*Taeniopygia guttata castanotis*). *Zeitschrift für Tierpsychologie* 69, 55–65.

Bishop, P.O., Jeremy, D. and Lance, J.W. (1953) The optic nerve. Properties of a central tract. *Journal of Physiology* 121, 415–432.

Blakemore, C. and Cooper, G.F. (1970) Development of the brain depends on the visual environment. *Nature* 228, 477–478.

Blakemore, C. and van Sluyters, R.C. (1974) Reversal of the physiological effects of monocular deprivation in kittens: further evidence for a sensitive period. *Journal of Physiology* 237, 195–216.

Blokhuis, H.J. (1984) Rest in poultry. *Applied Animal Behaviour Science* 12, 289–303.

Blokhuis, H.J. and Arkes, J.G. (1984) Some observations on the development of feather pecking in poultry. *Applied Animal Behaviour Science* 12, 145–157.

Blokhuis, H.J. and van der Haar, J.W. (1989) Effects of floor type during rearing and of beak trimming on ground pecking and feather pecking in laying hens. *Applied Animal Behaviour Science* 22, 359–369.

Blokhuis, H.J., van der Harr, J.W. and Koole, P.G. (1987) Effects of beak-trimming and floor type on feed consumption and body weight of pullets during rearing. *Poultry Science* 66, 623–625.

Boakes, R. and Panter, D. (1985) Secondary imprinting in the domestic chick blocked by previous exposure to a live hen. *Animal Behaviour* 33, 353–365.

Bode-Greuel, K.M. and Singer, W. (1989) The development of N-methyl-D-aspartate receptors in cat visual cortex. *Experimental Brain Research* 46, 197–204.

Bolhuis, J.J. (1991) Mechanisms of avian imprinting: a review. *Biological Reviews* 66, 303–345.

Bolhuis, J.J. and Bateson, P. (1990) The importance of being first: a primary effect of filial imprinting. *Animal Behaviour* 40, 472–483.

Bolhuis, J.J. and Trooster, W.J. (1988) Reversibility revisited: stimulus-dependent stability of filial preference in the chick. *Animal Behaviour* 36, 668–674.

Bolhuis, J.J. and van Kampen, H.S. (1992) An evaluation of auditory learning in filial imprinting. *Behaviour* 122, 195–230.

Bolhuis, J.J., Johnston, M.H. and Horn, G. (1985) Effects of early experience on the development of filial preferences in the domestic chick. *Developmental Psychobiology* 18, 299–308.

Bolhuis, J.J., McCabe, B.J. and Horn, G. (1986) Androgens and imprinting: differential effects of testosterone on filial preference in the domestic chick. *Behavioral Neuroscience* 100, 51–56.

Bolhuis, J.J., de Vos, G.J. and Kruijt, J.P. (1990) Filial imprinting and associative learning. *Quarterly Journal of Experimental Psychology* 42B, 313–329.

Bottjer, S.W., Schoonmaker, J.N. and Arnold, A.P. (1986) Auditory and hormonal stimulation interact to produce neural growth in adult canaries. *Journal of Neurobiology* 17, 605–612.

Bourne, R.C. and Stewart, M.G. (1985) Elevation in binding of ^{3}H-muscimol to high affinity GABA receptors in chick forebrain 24 hours after passive avoidance training. *Neuroscience Letters* S21, 573.

Boxer, M.I. and Stanford, D. (1984) The visual hyperstriatal region in the chick: potentials evoked by electrical stimuli. *Neuroscience Letters* 45, 323–328.

Boxer, M.I. and Stanford, D. (1985) Projections to the posterior visual hyperstriatal region a the chick: an HRP study. *Experimental Brain Research* 57, 494–498.

Bradford, C.M. and McCabe, B.J. (1992) An association between imprinting and spontaneous neuronal activity in the hyperstriatum ventrale of the anaesthetized domestic chick. *Journal of Physiology* 452, 238P.

Bradley, P.M. and Horn, G. (1981) Imprinting: a study of cholinergic receptor sites in parts of the chick brain. *Experimental Brain Research* 41, 121–123.

Bradley, P.M., Horn, G. and Bateson, P. (1981) Imprinting: an electron microscopic study of the chick hyperstriatum ventrale. *Experimental Brain Research* 41, 115–120.

Bradley, P.M., Davies, D.C. and Horn, G. (1985) Connections of the hyperstriatum ventrale of the domestic chick (*Gallus domesticus*). *Journal of Anatomy* 140, 577–589.

Bradshaw, J.L. and Rogers, L.J. (1992) *The Evolution of Lateral Asymmetries, Language. Tool Use and Intellect*. Academic Press, San Diego.

Braitenberg, V. and Kemali, M. (1970) Exceptions to bilateral symmetry in the epithalamus of lower vertebrates. *Journal of Comparative Neurology* 138, 137–146.

Braun, K., Bock, J. and Wolf, A. (1992) NMDA–mediated mechanisms in auditory filial imprinting in chicks. *Proceedings of Fifth Conference on the Neurobiology of Learning and Memory*. Irvine, CA, no. 113.

Breward, J. (1984) Cutaneous nociceptors in the chicken beak. *Journal of Physiology* 346, 56.

Breward, J. (1986) Cutaneous nociceptors in the amputated beak of the chicken. *Journal of Physiology* 3, 119.

Bronchti, G., Schönenberger, N., Welker, E. and Van der Loos, H. (1992) Barrelfield expansion after neonatal eye removal in mice. *Developmental Neuroscience* 3, 489–492.

Broom, D. (1968) Behavior of undisturbed 1- to 10-day-old chicks in different rearing conditions. *Developmental Psychobiology* 1, 287–295.

Broom, D.M. (1969) Reactions of chicks to visual changes during the first ten days after hatching. *Animal Behaviour* 17, 307–315.

Brown, M.W. and Horn, G. (1992) Neurones in the intermediate and medial part of the hyperstriatum ventrale (IMHV) of freely moving chicks respond to visual and/ or auditory stimuli. *Journal of Physiology* 452, 237P.

Bullock, S., Potter, J. and Rose, S.P.R. (1990) Identification of chick brain glycoproteins showing changed fucosylation rates after passive avoidance training. *Journal of Neurochemistry* 54, 135–142.

Bullock, S.P. and Rogers, L.J. (1985) Sex differences in the effects of testosterone and its metabolites on brain asymmetry for the control of copulation in young chicks. *Proceedings of the Australian Physiological and Pharmacological Society* 16, 235P.

Bullock, S.P. and Rogers, L.J. (1986) Glutamate-induced asymmetry in the sexual and aggressive behavior of young chickens. *Physiology, Biochemistry and Behavior* 24, 549–554.

Bullock, S.P. and Rogers, L.J. (1992) Hemispheric specialization for the control of copulation in the young chick and effects of 5α-dihydrotestosterone and 17β-oestradiol. *Behavioural Brain Research* 48, 9–14.

Burchuladze, R.A., Potter, J. and Rose, S.P.R. (1990) Memory formation in the chick depends on membrane-bound protein kinase C. *Brain Research* 535, 131–138.

Burne, T.H. and Rogers, L.J. (1994a) Olfactory imprinting in the domestic chick. *Proceedings of the Australian Neuroscience Society* 5, 231.

Burne, T.H. and Rogers, L.J. (1994b) Olfactory imprinting and learning. *Physiology and Behaviour* submitted.

Bursian, A.V. (1965) Primitive forms of photosensitivity at early stages of embryogenesis in the chick. *Journal of Evolutionary Biochemistry and Physiology* 1, 435–441.

Cannon, C.E. (1983) Descriptions of foraging behaviour of Eastern and Pale-headed Rosellas. *Bird Behaviour* 4, 63–70.

Catsicas, S., Catsicas, M. and Clarke, P.G.H. (1987) Long-distance intraretinal connections in birds. *Nature* 326, 186–187.

Cerella, J. (1986) Pigeons and perceptions. *Pattern Recognition* 19, 431–438.

Chaiken, M., Böhner, J. and Marler, P. (1993) Song acquisition in European starlings, *Sturnus vulgaris*: a comparison of the songs of live-tutored, tape-tutored, untutored, and wild-caught males. *Animal Behaviour* 46, 1079–1090.

Chaves, L.M., Hodos, W. and Güntürkün, O. (1993) Color-reversal learning: effects after lesions of thalamic visual structures in pigeons. *Visual Neuroscience* 10, 1099–1107.

Cherfas, J.J. (1977) Visual system activation in the chick: one-trial avoidance learning affected by duration and patterning of light exposure. *Behavioral Biology* 21, 52–65.

Cherfas, J.J. and Scott, A.M. (1981) Impermanent reversal of filial imprinting. *Animal Behaviour* 29, 301.

Cipolla-Neto, J., Horn, G. and McCabe, B.J. (1982) Hemispheric asymmetry and imprinting: the effect of sequential lesions of the hyperstriatum ventrale. *Experimental Brain Research* 48, 22–27.

Clarke, P.G.H. (1992) Neurone death in the developing avian isthmo-optic nucleus, and its relation to the establishment of functional circuitry. *Journal of Neurobiology* 23, 1140–1158.

Clayton, N.S. (1993) Lateralization and unilateral transfer of spatial memory in marsh tits. *Journal of Comparative Physiology A* 171, 799–806.

Clayton, N.S. (1994) Marsh tits (*Parus palustris*) use tools to store food. *British Birds* submitted.

Clayton, N.S. and Krebs, J.R. (1993) Lateralization in Paridae: comparison of a storing and a non-storing species on a one-trial associative memory task. *Journal of Comparative Physiology A* 171, 807–815.

Clayton, N.S. and Krebs, J.R. (1994a) Hippocampal growth and attrition in birds affected by experience. *Proceedings of the National Academy of Science USA* 91, 7410–7414.

Clayton, N.S. and Krebs, J.R. (1994b) Lateralization and unilateral transfer of spatial memory in marsh tits: Are two eyes better than one? *Journal of Comparative Physiology A* 174(6), 769–773.

Clifton, P.G. and Andrew, R.J. (1983) The role of stimulus and colour in the elicitation of testosterone-facilitated aggressive and sexual responses in the domestic chicks. *Animal Behaviour* 31, 878–901.

Cohen, J. (1981) Olfaction and parental behavior in ring doves. *Biochemical Systematics and Ecology* 9, 351–354.

Collias, N.E. (1952) The development of social behavior in birds. *Auk* 69, 127–159.

Collins, J.W. and Siegel, P.B. (1987) Human handling, flock size and responses to an *E. coli* challenge in young chickens. *Applied Animal Behavior Science* 19, 183–188.

Collins, R.L. (1985) On the inheritance of direction and degree of asymmetry. In: Glick, S.D. (ed.) *Cerebral Lateralization in Nonhuman Species.* Academic Press, Orlando and London, pp. 41–71.

Cook, R.G. (1993) Gestalt contributions to visual texture discriminations by pigeons. In: Zentall, T.R. (ed.) *Animal Cognition.* Lawrence Erlbaum, Hillsdale, NJ, pp. 251–269.

Cook, S.E. (1993) Retention of primary preferences after secondary filial imprinting. *Animal Behaviour* 46, 405–407.

Corner M.A. and Bakhuis, W.L. (1969) Developmental patterns in the central nervous system of birds. V. Cerebral electrical activity, forebrain function and behavior in the chick at the time of hatching. *Brain Research* 13, 541–555.

Corner, M.A. and Schadé, J.P. (1967) Developmental patterns in the central nervous system of birds. IV. Cellular and molecular bases of functional activity. *Progress in Brain Research* 26, 237–250.

Corner, M.A., Bakhuis, W.L. and van Wingerden, C. (1973) Sleep and wakefulness during early life in the domestic chicken, and their relationship to hatching and embryonic motility. In: Gottlieb, G. (ed.) *Behavioral Embryology.* Volume 1, Academic Press, New York, pp. 245–279.

Cotman, C.W. and Iversen, L.L. (1987) Excitatory amino acids in the brain – focus on NMDA receptors. *Trends in Neurosciences* 7, 263–266.

Coulombre, A.J. (1955) Correlations of structural and biochemical changes in the developing retina of the chick. *American Journal of Anatomy* 96, 153–187.

Cowan, P.J. (1974) Selective responses to the parental calls of different individual hens by young *Gallus gallus*: auditory discrimination learning versus auditory imprinting. *Behavioural Biology* 10, 541–545.

Cowan, P.J. and Evans, R.M. (1974) Calls of different individual hens and the parental control of feeding behavior in young *Gallus gallus. Journal of Experimental Zoology* 188, 353–360.

Croucher, S.J. and Tickle,C. (1989) Characterization of epithelial domains in the nasal passages of chick embryos: spatial and temporal mapping of a range of extracellular matrix and cell surface molecules during development of the nasal placode. *Development* 106, 493–509.

Crowe, S.F., Ng, K.T. and Gibbs, M.E. (1990) Memory consolidation of weak training experiences by hormonal treatments. *Pharmacology, Biochemistry and Behaviour* 37, 729–734.

Croze, H. (1970) Search image in carrion crows. *Zeitschrift für Tierpsychologie*, Suppl. 5, 1–85.

Cuénod, M. and Streit, P. (1980) Amino acid transmitters and local circuitory in the optic tectum. In: Schmidt, F.O. and Warden, F.G. (eds) *The Neurosciences*. MIT Press, Cambridge, MA, pp. 989–1004.

Cynx, J. and Nottebohm, F. (1992) Testosterone facilitates some conspecific song discriminations in castrated zebra finches (*Taeniopygia guttata*). *Proceedings of the National Academy of Science USA* 89, 1376–1378.

Cynx, J., Williams, H. and Nottebohm, F. (1992) Hemispheric differences in avian song discrimination. *Proceedings of the National Academy Science* 89, 1372–1375.

Davey, J.E. and Horn, G. (1991) The development of hemispheric asymmetries in neuronal activity in the domestic chick after visual experience. *Behavioural Brain Research* 45, 81–86.

Davies, D., Horn, C. and McCabe, B.J. (1983) Changes in telencephalic catecholamine levels in the domestic chick. Effects of age and visual experience. *Developmental Brain Research* 10, 251–255.

Davies, D., Horn, C. and McCabe, B.J. (1985) Noradrenaline and learning: the effects of the neurotoxin DSP4 on imprinting in the domestic chick. *Behavioral Neuroscience* 100, 51–56.

Davies, M.N.O. and Green, P.R. (1991) Footedness in pigeons, or simply sleight of foot? *Animal Behaviour* 42, 311–312.

Davis, S.J. and Fischer, G.J. (1978) Chick colour preferences are altered by cold stress: colour pecking and approach preferences are the same. *Animal Behaviour* 26, 259–264.

Dawkins, M.S. (1980) *Animal Suffering: The Science of Animal Welfare*, Chapman and Hall, London.

Dawkins, M.S. and Beardsley, J. (1986) Reinforcing properties of access to litter in hens. *Applied Animal Behaviour Science* 15, 351–364.

Dawkins, R. (1968) The ontogeny of a pecking preference in domestic chicks. *Zeitschrift für Tierpsychologie* 25, 170–186.

Dawkins, R. (1969) A threshold model of choice behaviour. *Animal Behaviour* 17, 120–133.

Delius, J.D. (1985) Cognitive processes in pigeons. In: D'Ydelvalle, G. (ed.) *Cognition, Information Processing and Motivation*. Elsevier, Amsterdam, pp. 3–18.

Delius, J.D. (1990) Sapient sauropsids and hollering hominids. In: Koch, W.A. (ed.) *Geneses of Language*. Brockmeyer, Bochum, pp. 1–29.

Denbow, D.M. (1991) Induction of food intake by a GABAergic mechanism in the turkey. *Physiology and Behavior* 49, 485–488.

Denenberg, V.H. (1981) Hemispheric laterality in animals and the effects of early experience. *Behavioural Brain Science* 4, 1–49.

Denenberg, V.H. (1984) Behavioural asymmetry. In: Geschwind, N. and Galaburda, A.M. (eds) *Cerebral Dominance: The Biological Foundations*. Harvard University Press, Cambridge, MA, pp. 114–133.

Denenberg, V.H., Zeidner, L., Rosen, G.D., Hofmann, M., Garbanati, J.A., Sherman, G.F. and Yutzey, D.A. (1981) Stimulation in infancy facilitates interhemispheric communication in the rabbit. *Developmental Brain Research* 1, 165–169.

Denenberg, V.H., Hofmann, M.J., Rosen, G.D. and Yutzey, D.A. (1984) Cerebral asymmetry and behavioral laterality: some psychobiological considerations. In: Fox, N.A. and Davidson, R.J. (eds) *The Psychobiology of Affective Development*. Erlbaum, Hillsdale, NJ, pp. 77–117.

Denenberg, V.H., Fitch, R.H., Schrott, L.M., Cowell, P.E. and Waters, N.S. (1991) Corpus callosum: interactive effects of infantile handling and testosterone in the rat. *Behavioral Neuroscience* 105, 562–566.

Deng, C. and Wang, B. (1992) Overlap of somatic and visual response areas in the Wulst of pigeons. *Brain Research* 582, 320–323.

de Vos, G.J. and Bolhuis, J.J. (1990) An investigation into blocking of filial imprinting in the chick during exposure to a compound stimulus. *Quarterly Journal of Experimental Psychology* 42B, 289–312.

de Vos, G.J. and van Kampen, H.S. (1993) Effects of primary imprinting on the subsequent development of secondary filial attachments in the chick. *Behaviour* 125, 245–263.

Dewsbury, D.A. (1982) Dominance rank, copulatory behavior, and differential reproduction. *Quarterly Review of Biology* 57, 135–159.

Dewsbury, D.A. (1990) Fathers and sons: genetic factors and social dominance in deer mice, *Peromyscus maniculatus*. *Animal Behaviour* 39, 284–289.

Dharmaretnam, M. and Andrew, R.J. (1994) Age- and stimulus-specific effects on use of right and left eyes by the domestic chick. Submitted.

Diamond, M.C. (1985) Rat forebrain morphology: Right–left: Male–female: Young–old: Enriched–impoverished. In: Glick, S.D. (ed.) *Cerebral Lateralization in Nonhuman Species*, Academic Press, New York, pp. 73–87.

Diamond, M.E., Armstrong-Jones, M. and Ebner, F.F. (1993) Experience-dependent plasticity in adult rat barrel cortex. *Proceedings of the National Academy of the Sciences, USA* 90, 2082–2086.

Dimond, S.J. (1968) Effects of photic stimulation before hatching on the development of fear in chicks. *Journal of Comparative and Physiological Psychology* 65, 320–324.

Dimond, S.J. and Adam, J.H. (1972) Approach behaviour and embryonic visual experience in chicks: studies of the rate of visual flicker. *Animal Behaviour* 20, 413–420.

Dmitrieva, L.P. and Gottlieb, G. (1992) Development of brainstem auditory pathway in mallard duck embryos and hatchlings. *Journal of Comparative Physiology A* 171, 665–671.

Dörsam, R., Wallhäusser-Franke, E., Steffen, H. and Scheich, H. (1991) Morphology of projection neurons in chick MNH. *Abstracts of Avian Learning and Plasticity Conference*. Milton Keynes, UK, p. 55.

Dubbeldam, J.L. (1991) The avian and mammalian forebrain: correspondences and differences. In: Andrew, R.J. (ed.) *Neural and Behavioural Plasticity: The Use of the Domestic Chick as a Model*. Oxford University Press, Oxford, pp. 536–554.

Ducker, G., Luscher, C. and Schultz, P. (1986) Problemlöseverhalten von Stieglitzen (*Carduelis carduelis*) bei 'manipulativen' Aufgaben. *Zoologische Beiträge* 23, 377–412.

Duyck, I. and Duyck, J. (1984) Koolmes, Parus major, gebruikt instrument bij het voedselzoeken. *Wielewaal* 50, 416.

Dyer, A.B., Lickliter, R. and Gottlieb, G. (1989) Maternal and peer imprinting in mallard ducklings under experimentally simulated natural social conditions. *Developmental Psychobiology* 22, 463–475.

Eccles, J.C. (1992) Evolution of consciousness. *Proceedings of the National Academy of Science* 89, 7320–7324.

Ehret, G. (1987) Left hemisphere advantage in the mouse brain for recognizing ultrasonic communication calls. *Nature* 325, 249–251.

Ehrlich, D. (1981) Regional specialisation of the chick retina as revealed by the size and density of neurones in the ganglion cells layer. *Journal of Comparative Neurology* 195, 643–657.

Ehrlich, D. and Mark, R. (1984) An atlas of the primary visual projections in the brain of the chick *Gallus gallus*. *Journal of Comparative Neurology* 223, 592–610.

Elberger, A.J. (1982) The functional role of the corpus callosum in the developing visual system: a review. *Progress in Neurobiology* 18, 15–79.

Emmerton, J. and Delius, D. (1993) Beyond Sensation: visual cognition in pigeons. In: Zeigler, H.P. and Bischof, H-J. (eds) *Vision, Brain and Behavior in Birds*. MIT Press, Cambridge, MA, pp. 377–390.

Engbretson, G.A., Reiner, A. and Brecha, N. (1981) Habenular asymmetry and the central connections of the parietal eye of the lizard. *Journal of Comparative Neurology* 198, 155–165.

Engelage, J. and Bischof, H-J. (1993) The organisation of the tectofugal pathway in birds: a comparative review. In: Zeigler, H.P. and Bischof, H-J. (eds) *Vision, Brain, and Behavior in Birds*. MIT Press, Cambridge, MA, pp. 137–158.

Epstein, R., Kirshnit, C.E., Lanza, R.P. and Rubin, L.C. (1984) 'Insight' in the pigeon: antecedents and determinants of an intelligent performance. *Nature* 308, 61–62.

Epstein, R., Lanza, R.P. and Skinner, B.F. (1981) 'Self-awareness' in the pigeon. *Science* 212, 695–696.

Erichsen, J., Bingman, V. and Krebs, J. (1991) The distribution of neuropeptides in the hippocampal region of the pigeon (*Columba livia*): a basis of regional subdivisions. *Journal of Comparative Neurology* 245, 478–492.

Evans, C.S. and Marler, P. (1991) On the use of video images as social stimuli in birds: audience effects on alarm calling. *Animal Behaviour* 41, 17–26.

Evans, C.S. and Marler, P. (1992) Female appearance as a factor in the responsiveness of male chickens during anti-predator behaviour and courtship. *Animal Behaviour* 43, 137–145.

Evans, C.S. and Marler, P. (1994) Food calling and audience effects in male chickens, *Gallus gallus*: their relationships to food availability, courtship and social facilitation. *Animal Behaviour* 47, 1159–1170.

Evans, C.S., Evans, L. and Marler, P. (1993a) On the meaning of alarm calls: functional references in an avian vocal system. *Animal Behaviour* 46, 23–28.

Evans, C.S., Macedonia, J.M. and Marler, P. (1993b) Effects of apparent size and speed on the response of chickens, *Gallus gallus*, to computer-generated simulation of aerial predators. *Animal Behaviour* 46, 1–11.

Evans, R.M. (1975) Stimulus intensity and acoustical communication in young domestic chicks. *Behaviour* 55, 73–80.

Evans, R.M. (1982) The development of learned auditory discriminations in the context of post-natal filial imprinting in young precocial birds. *Bird Behaviour* 4, 1–6.

Fagg, G.E. and Foster, A.C. (1983) Amino acid neurotransmitters and their pathways in the mammalian central nervous system. *Neuroscience* 9, 701–719.

Fagg, G.E. and Matus, A. (1984) Selective association of *N*-methyl aspartate and quisquilate types of ʟ-glutamate receptor with postsynaptic densities. *Proceedings of the National Academy of the Sciences USA* 81, 6876–6880.

Fält, B. (1981) Development of responsiveness to the individual maternal 'clucking' by domestic chicks (*Gallus gallus domesticus*). *Behavioural Processes* 6, 303–317.

Faure, J.M. and Jones, R.B. (1982) Effects of sex, strain and type of perch on perching behaviour in the domestic fowl. *Applied Animal Ethology* 8, 281–293.

Fisher, L.J. (1983) Development of retinal synaptic arrays in mouse, chicken and Xenopus: a compative study. In: Sheffield, J.B. and Hilfer, S.R. (eds) *Cellular Communication During Ocular Development*. Springer-Verlag, New York, pp. 15–30.

Fiszer de Plazas, S., Floves, V. and Rios, H. (1986) GABA receptors in the developing chick visual system: influence of light and darkness. In: Racagni, G. and Donoso, A. (eds) *GABA and Neuroendocrine Function*. Raven Press, New York, pp. 25–38.

Fiszer de Plazas, S., Conterjnic, D. and Flores, V. (1991) Effect of a simple visual pattern on the early postnatal development of GABA receptor sites in the chick optic lobe. *International Journal of Developmental Neuroscience* 9, 195–201.

Fitch, R.H., Cowell, P.E., Schrott, L.M. and Denenberg, V.H. (1991) Corpus callosum: ovarian hormones and feminization. *Brain Research* 542, 313–317.

Foelix, R.F. and Oppenheim, R.W. (1973) Synaptogenesis in the avian embryo: ultrastructure and possible behavioral correlates. In: Gottlieb, G. (ed.) *Behavioral Embryology*, Volume 1. Academic Press, New York, pp. 103–139.

Fontanesi, G., Casini, G., Ciocchetti, A. and Bagnoli, P. (1993) Development, plasticity, and differential organisation of parallel processing of visual information in birds. In: Zeigler, H.P. and Bischof, H-J. (eds) *Vision, Brain and Behavior in Birds*. MIT Press, Cambridge, MA, pp. 195–205.

Foster, A.C. and Wong, E.H.F. (1987) The novel anticonvulsant MK-801 binds to the activated state of the *N*-methyl-ᴅ-aspartate receptor in rat brain. *British Journal of Pharmacology* 91, 403–409.

Freeman, B.M. (1965) The relationship between oxygen consumption, body temperature and surface area in the hatching and young chick. *British Poultry Science* 6, 67–72.

Freeman, B.M. and Vince, M.A. (1974) *Development of the Avian Embryo*. Chapman and Hall, London.

Friedman, H. and Davis, M. (1938) 'Left-handedness' in parrots. *Auk* 80, 478–480.

Friend, T.H. and Polan, C.E. (1978) Competitive order as a measure of social dominance in dairy cattle. *Applied Animal Ethology* 4, 61–70.

Fritzsch, B., Crapou de Caprona, M.-D. and Clarke, P.G.H. (1990) Development of two morphological types of retinopetal fibres in chick embryos, as shown by diffusion along axons of a carbocyanine dye in the fixed retina. *Journal of Comparative Neurology* 300, 405–421.

Gahr, M., Güttinger, H-R. and Kroodsma, D.E. (1993) Estrogen receptors in the avian brain: survey reveals general distribution and forebrain areas unique to songbirds. *Journal of Comparative Neurology* 327, 112–122.

Galli-Resta, L., Ensini, M., Fusco, E., Gravina, A. and Margheritti, B. (1993) Afferent spontaneous electrical activity promotes the survival of target cells in the developing retinotectal system of the rat. *Journal of Neuroscience* 13, 243–250.

Gallup, G.G. Jr (1974) Animal hypnosis: factual status of a fictional concept. *Psychological Bulletin* 81, 836–853.

Ganchrow, D. and Ganchrow, J. R. (1985) Number and distribution of taste buds in the oral cavity of hatching chicks. *Physiology and Behaviour* 34, 889–894.

Garcia-Austt, E. and Patetta-Queirolo, M.A. (1961) Electroretinogram of the chick embryo. I. Onset and development. *Acta Neurologia Latinoamerica* 7, 179–189.

Garcia-Austt, E. and Patetta-Quierolo, M.A. (1961) Electroretinogram of the chick embryo. II. Influence of adaptations, flicker frequency and wavelength. *Acta Neurologia Latinoamerica* 7, 269–288.

Gardino, P.F., Santos, R.M. and Hokoç, J.N. (1993). Histogenesis and topographical distribution of tyrosine hydroxylase immunoreactive amacrine cells in the developing chick retina. *Developmental Brain Research* 72, 226–236.

Gaston, K.E. (1984) Interocular transfer of pattern discrimination learning in chicks. *Brain Research* 310, 213–221.

Gaston, K.E. and Gaston, M.G. (1984) Unilateral memory after binocular discrimination training: left hemisphere dominance in the chick? *Brain Research* 303, 190–193.

Gentle, M.J. (1971) Taste and its importance to the domestic chicken. *British Poultry Science* 12, 77–86.

Gentle, M.J. (1972) Taste preference in the chicken (*Gallus domesticus* L.) *British Poultry Science* 13, 141–155.

Gentle, M.J. (1975) Gustatory behaviour of the chicken and other birds. In: Wright, P., Caryl, P.G. and Vowles, D.M. (1975) *Neural and Endocrine Aspects of Behaviour in Birds.* Elsevier, Amsterdam, pp. 305–318.

Gentle, M.J., Hughes, B.O. and Hubrecht, R.C. (1982) The effect of beak trimming on food intake, feeding behaviour and body weight in adult hens. *Applied Animal Ethology* 8, 147–159.

Geschwind, N. and Galaburda, A.M. (1987) *Cerebral Lateralization: Biological Mechanisms, Associations, and Pathology.* MIT Press, Cambridge, MA.

Gibbs, M.E. (1991) Behavioral and pharmacological unravelling of memory formation. *Neurochemical Research* 16, 715–726.

Gibbs, M.E. and Ng, K.T. (1977) Psychobiology of memory: towards a model of memory formation. *Biobehaviour Reviews* 1, 113–136.

Gibbs, M.E., Ng, K.T. and Crowe, S. (1991) Hormones and the timing of phases of memory formation. In: Andrew, R.J. (ed.) *Neural and Behavioural Plasticity: The Use of the Domestic Chick as a Model.* Oxford University Press, Oxford, pp. 440–455.

Gibson, K.R. and Ingold, T. (1993) *Tools, Language and Cognition in Human Evolution.* Cambridge University Press, Cambridge.

Girard, H. (1973) Arterial pressure in the chick embryo. *American Journal of Physiology* 224, 454–460.

Glick, S.D. (1985) Heritable differences in turning behavior of rats. *Life Sciences* 36, 499–503.

Gold, P. (1969) Effects of prehatching stimulation on growth and behavior in the domestic chick. *American Zoologist* 9, 1074.

Goldman, S. and Nottebohm, F. (1983) Neuronal production, migration, and differentiation in a vocal control nucleus of the adult female canary brain.

Proceedings of the National Academy of Sciences of the USA 80, 2390–2394.

Gonzales, C.B., Charreau, E.H., Aragones, A., Lantos, C.P. and Follett, B.K. (1987) The autogenesis of reproductive hormones in the female embryo of the domestic fowl. *General and Comparative Endocrinology* 68, 369–374.

Goodwin, E.H. and Hess, E.H. (1969) Innate visual form preferences in the pecking behaviour of young chickens. *Behaviour* 34, 223–237.

Gottier, R.F. (1968) The dominance–submissive hierarchy in the social behaviour of the domestic chicken. *Journal of Genetic Psychology* 112, 205–226.

Gottlieb, G. (1965a) Prenatal auditory sensitivity in chickens and ducks. *Science* 147, 1596–1598.

Gottlieb, G. (1965b) Imprinting in relation to parental and species identification by avian neonates. *Journal of Comparative and Physiological Psychology* 59, 345–356.

Gottlieb, G. (1968). Prenatal behavior in birds. *Quarterly Review of Biology* 43, 148–174.

Gottlieb, G. (1971a) *Development of Species Identification in Birds: An Inquiry into the Prenatal Determinants of Perception.* University of Chicago Press, Chicago.

Gottlieb, G. (1971b) Ontogenesis of sensory function in birds and mammals. In: Tobach, E., Aronson, LA. and Shaw, E. (eds) *The Biopsychology of Development.* Academic Press, New York, pp. 67–128.

Gottlieb, G. (1973) Introduction to behavioral embryology. In: Gottlieb, G. (ed.) *Behavioral Embryology,* Volume 1. Academic Press, New York, pp. 3–45.

Gottlieb, G. (1976) Conceptions of prenatal development: behavioral embryology. *Psychological Review* 83, 215–234.

Gottlieb, G. (1978) Development of species identification in ducklings: IV. Change in species-specific perception caused by auditory deprivation. *Journal of Comparative and Physiological Psychology* 92, 375–387.

Gottlieb, G. (1979) Development of species identification in ducklings: V. Perceptual differentiation in the embryo. *Journal of Comparative and Physiological Psychology* 93, 831–854.

Gottlieb, G. and Klopfer, P.H. (1962) The relation of development age to auditory and visual imprinting. *Journal of Comparative and Physiological Psychology* 55, 821–826.

Gottlieb, G. and Kuo, Z-Y. (1965) Development of behavior in the duck embryo. *Journal of Comparative and Physiological Psychology* 59, 183–188.

Gottlieb, G. and Simner, M.L. (1969) Auditory versus visual flicker in directing the approach response of domestic chicks. *Journal of Comparative and Physiological Psychology* 67, 58–63.

Gottlieb, G. and Vandenbergh, J.G. (1968) Ontogeny of vocalisation in duck and chick embryos. *Journal of Experimental Zoology* 168, 307–326.

Graves, H.B. (1973) Early social responses in *Gallus*: a functional analysis. *Science* 182, 937–938.

Gravielle, M.C., Flores, V. and De Plazas, S.F. (1992) The postnatal development of benzodiazepine receptor sites in the chick optic lobe is modulated by environmental lighting. *Neurochemistry International* 20, 257–262.

Gray, L. and Rubel, E.W. (1985) Development of absolute thresholds in chickens. *Journal of the Acoustic Society of America* 77, 1162–1172.

Greuel, J., Luhmann, H.J. and Singer, W. (1988) Pharmacological induction of use-dependent receptive field modifications in the visual cortex. *Science* 242, 74–77.

Grier, J.B., Counter, S.A. and Shearer, W.M. (1967) Prenatal auditory imprinting in chickens. *Science* 155, 1692–1693.

Gross, W.B. and Siegel, P.B. (1982) Socialization as a factor in resistance to infection, feed efficiency, and response to antigen in chickens. *American Journal of Veterinary Research* 43, 2010–2012.

Gu, Q., Bear, M. and Singer W. (1989) Blockade of NMDA-receptors prevents ocularity changes in kitten visual cortex after reversed monocular deprivation. *Developmental Brain Research* 47, 281–288.

Guhl, A.M. (1958) The development of social organisation in the domestic chick. *Animal Behaviour* 6, 92–111.

Guilford, T., Nicol, C., Rothschild, M. and Moore, B.P. (1987) The biological roles of pyrazines: evidence for a warning odour function. *Biological Journal of the Linnean Society* 31, 113–128.

Güntürkün, O. (1985) Lateralization of visually controlled behaviour in pigeons. *Physiology and Behaviour* 34, 575–577.

Güntürkün, O. (1990) Embryonale Orientierung als Ontogenetischer Auslöser für Visuelle Lateralization. In: Frey, D. (ed.) *Bericht über den 37. Kongreß der Deutschen Gesellschaft für Psychologie in Kiel 1990*, Band 1. Verlag für Psychologie, pp. 51–52.

Güntürkün, O. (1993) The ontogeny of visual lateralization in pigeons. *German Journal of Psychology* 17, 276–287.

Güntürkün, O. and Bohringer, P.J. (1987) Lateralization reversal after intertectal commissurotomy in the pigeon. *Brain Research* 408, 1–5.

Güntürkün, O. and Hahmann, U. (1994) Visual acuity and hemispheric asymmetries in pigeons. *Behavioural Brain Research* 60, 171–175.

Güntürkün, O. and Hoferichter, H. (1985) Neglect after section of a left telencephalotectal tract in pigeons. *Behavioural Brain Research* 18, 1–9.

Güntürkün, O. and Kesch, S. (1987) Visual lateralization during feeding in pigeons. *Behavioural Neuroscience* 101, 433–435.

Güntürkün, O. and Kischkel, K-F. (1992) Is visual lateralization in pigeons sex-dependent? *Behavioural Brain Research* 47, 83–87.

Güntürkün, O., Kesch, S. and Delius, J.D. (1988) Absence of footedness in domestic pigeons. *Animal Behaviour* 36, 602–604.

Güntürkün, O., Emmerton, J. and Delius, J.D. (1989) Neural aysmmetries and visual behavior in birds. In: Lüttgau, H.C. and Necker, R. (eds) *Biological Signal Processing*. FRG, Deutsche Forschungsgemeinschaft, VCH Verlagsgesellschaft, Weinheim, pp. 122–145.

Gurusinghe, C.J. and Ehrlich, D. (1985) Sex-dependent structural asymmetry of the medial habenular nucleus of the chicken brain. *Cellular and Tissue Research* 240, 149–152.

Guyomarc'h J.-C. (1966) Les emissions sonores du poussin domestique, leur place dans le camportement normal. *Zeitschrift für Tierpsychologie* 23, 141–160.

Guyomarc'h, J.-C. (1972) Les bases ontogénétiques de l'attractivité du gloussement maternel chez la poule domestique. *Revue de Comportement Animal* 6, 79–94.

Guyomarc'h, J.-C. (1974) L'empreinte auditive prenatale chez le poussin domestique. *Revue de Comportement Animal* 8, 3–6.

Guyomarc'h, J.-C. (1975) Les cycles d'activité d'une couvée naturelle de poussins et leur coordination. *Behaviour* 53, 31–75.

Gyger, M., Karakashian, S., Dufty, A.M. and Marler, P. (1988) Alarm signals in birds: the role of testosterone. *Hormones and Behaviour* 22, 305–314.

Gyger, M., Karakashian, S.J. and Marler, P. (1986) Avian alarm calling: is there an audience effect? *Animal Behaviour* 34, 1570–1572.

Gyger, M., Marler, P. and Pickert, R. (1987) Semantics of an avian alarm call system: the male domestic fowl, *Gallus domesticus. Behaviour* 102, 15–40.

Hale, C. and Green, L. (1988) Effects of early ingestional experiences on the acquisition of appropriate food selection by young chicks. *Animal Behaviour* 36, 211–224.

Hale, E.B. (1969) Domestication and the evolution of behaviour. In: Hafez, E.S.E. (ed.) *The Behaviour of Domestic Animals*, 2nd edition. Ballière, Tindall and Cassell, London, pp. 22–42.

Hambley, J.W. and Rogers, L.J. (1979) Retarded learning induced by intracerebral administration of amino acids in the neonatal chick. *Neuroscience* 4, 677–684.

Hamburger, V. (1963) Ontogeny of behavior and its structural basis. In: Richter, D. (ed.) *Comparative Neurochemistry*. Pergamon, New York, pp. 21–34.

Hamburger, V. (1973) Anatomical and physiological basis of embryonic motility in birds and mammals. In: Gottlieb, G. (ed.) *Behavioral Embryology*, Volume 1. Academic Press, New York, pp. 51–76.

Hamburger, V. and Balaban, M. (1963) Observations and experiments on spontaneous rhythmical behavior in the chick embryo. *Developmental Biology* 7, 533–545.

Hamburger, V. and Hamilton, H.L. (1951) A series of normal stages in the development of the chick embryo. *Journal of Morphology* 88, 49–92. Reprinted (1992) in *Developmental Dynamics* 195, 231–272.

Hamburger, V. and Oppenheim, R. (1967) Prehatching motility and hatching behavior in the chick. *Journal of Experimental Zoology* 166, 171–204.

Hamburger, V., Balaban, M., Oppenheim, R. and Wenger, E. (1965) Periodic motility of normal and spinal chick embryos between 8 and 17 days of incubation. *Journal of Experimental Zoology* 159, 1–14.

Hamburger, V., Wenger, E. and Oppenheim, R. (1966) Motility in the chick embryo in the absence of sensory input. *Journal of Experimental Zoology* 162, 133–160.

Harris, L.J. (1989) Footedness in parrots: three centuries of research, theory, and mere surmise. *Canadian Journal of Psychology* 43, 369–396.

Harrison, J.R. (1951) *In vitro* analysis of differentiation of retinal pigment in the developing chick embryo. *Journal of Experimental Zoology* 118, 209–241.

Hartley, R.J. and Suthers, R.A. (1990) Lateralization of syringeal function during song production in the canary. *Journal of Neurobiology* 21, 1236–1248.

Hauser, M.D. (1993) Right hemisphere dominance for the production of facial expression in monkeys. *Science* 261, 475–477.

Haywood, J., Hambley, J.W. and Rose, S.P.R. (1975) Effects of exposure to an imprinting stimulus on the activity of enzymes involved in acetylcholine metabolism in chick brain. *Brain Research* 92, 219–225.

Healy, S.D. and Krebs, J.R. (1991) Hippocampal volume and migration in passerine birds. *Naturwissenschaften* 78, 424–426.

Healy, S.D. and Krebs, J.R. (1992) Food storing and the hippocampus in corvids: amount and volume are correlated. *Proceedings of the Royal Society of London B* 248, 241–245.

Healy, S.D. and Krebs, J.R. (1993) Development of hippocampal specialisation in a food-storing bird. *Behavioural Brain Research* 53, 127–131.

Heath, J.W. and Glenfield, P.J. and Rostas, J.A.P. (1992) Structural maturation of synapses in the rat superior cervical ganglion continues beyond four weeks of age. *Neuroscience Letters* 142, 17–21.

Heaton, M.B. (1976) Developing visual function in the red jungle fowl embryo. *Journal of Comparative and Physiological Psychology* 90, 53–56.

Heaton, M.B. and Harth, M.S. (1974) Developing visual function in the pigeon embryo with comparative reference to other avian species. *Journal of Comparative and Physiological Psychology* 86, 151–156.

Heil, P., Langner, G. and Scheich, H. (1992) Processing of frequency-modulated stimuli in the chick auditory cortex analogue: evidence for topographic representations and possible mechanisms of rate and directional sensitivity. *Journal of Comparative Physiology A* 171, 583–600.

Herrmann, K. and Bischof, H-J. (1986) Delayed development of song control nuclei in the zebra finch is related to behavioral development. *Journal of Comparative Neurology* 245, 167–175.

Herrmann, K. and Bischof, H-J. (1993) Development of the tectofugal visual system of normal and deprived zebra finches. In: Zeigler, H.P. and Bischof, H-J. (eds) *Vision, Brain and Behavior in Birds*. MIT Press, Cambridge, MA, pp. 207–226.

Herrnstein, R.J. (1982) Stimuli and the texture of experience. *Neuroscience and Behavioral Reviews* 6, 105–117.

Hess, E.H. (1956) Natural preferences of chicks and ducks for objects of different colours. *Psychological Reports* 2, 477–483.

Hess, E.H. (1959a) Imprinting. *Science* 130, 133–141.

Hess, E.H. (1959b) Two conditions limiting critical age for imprinting. *Journal of Comparative and Physiological Psychology* 52, 515–518.

Hinde, R. (1955) The following response of moorhens and coots. *British Journal of Animal Behaviour* 3, 121–122.

Hodos, W. (1969) Color discrimination deficits after lesions of nucleus rotundus in pigeons. *Brain, Behavior and Evolution* 2, 185–200.

Hodos, W. (1976) Vision and the visual system. In: Sprague, J.M. and Epstein, A.N. (eds) *Progress in Psychobiology and Physiological Psychology*, vol. 6, Academic Press, New York, pp. 29–62.

Hodos, W. and Erichsen, J.T. (1990) Lower-field myopia in birds: an adaptation that keeps the ground in focus. *Vision Research* 30, 653–657.

Hodos, W. and Karten, H.J. (1970) Visual intensity and pattern discrimination deficits after lesions of ectostriatum in pigeons. *Journal of Comparative Neurology* 140, 53–68.

Hodos, W., Macko, K.A. and Bessette, B.B. (1984) Near-field acuity changes after visual system lesions in pigeons. II. Telencephalon. *Behavioural Brain Research* 13, 15–30.

Hodos, W., Weiss, S.R.B. and Bessette, B.B. (1986) Size-threshold changes after lesions of the visual telencephalon in pigeons. *Behavioural Brain Research* 21, 203–214.

Hogan, J.A. (1966) An experimental study of conflict and fear: an analysis of behaviour of young chicks towards a mealworm. *Behaviour* 27, 273–289.

Hogan, J.A. (1971a) Development of food recognition in young chicks; IV associative and non-associative effects of experience. *Journal of Comparative Physiology and Psychology* 914, 839–850.

Hogan, J.A. (1971b) The development of a hunger system in young chicks. *Behaviour* 39, 128–201.

Hogan, J.A. (1973) The development of food recognition in young chicks: I. Maturation and nutrition. *Journal of Comparative and Physiological Psychology* 83, 355–366.

Hollard, V.D. and Delius, J.D. (1982) Rotational invariance in visual pattern recognition by pigeons and humans. *Science* 218, 804–806.

Horn, G. (1985) *Imprinting, Memory, and the Brain*. Clarendon Press, Oxford.

Horn, G. (1990) Neural bases of recognition memory investigated through the analysis of imprinting. *Philosophical Transactions of the Royal Society of London B* 329, 133–142.

Horn, G. (1991) Imprinting and recognition memory; a review of neural mechanisms. In: Andrew, R.J. (ed.) *Neural and Behavioral Plasticity: The Domestic Chick as a Model*. Oxford University Press, Oxford, pp. 219–261.

Horn, G. and Johnson, M.H. (1989) Memory systems in the chick: dissociations and neuronal analysis. *Neuropsychologia* 27, 1–22.

Horn, G. and McCabe, B.J. (1990) The time course of N-methyl-D-aspartate (NMDA) receptor binding in the chick brain after imprinting. *Journal of Physiolology* 423, 92P.

Horn, G., McCabe, B.J. and Bateson, P.P.G. (1979) An autoradiographic study of the chick brain after imprinting. *Brain Research* 168, 361–373.

Horn, G., Bradley, P. and McCabe, B.J. (1985) Changes in the structure of synapses associated with learning. *Journal of Neuroscience* 5, 3161–3168.

Howard, K.J., Rogers, L.J. and Boura, A.L.A. (1980) Functional lateralization of the chicken forebrain revealed by use of intracranial glutamate. *Brain Research* 188, 369–382.

Hughes, B.O. and Duncan, I.J.H. (1972) The influence of strain and environmental factors upon feather pecking and cannibalism in fowls. *British Poultry Science* 13, 525–547.

Hunt, E.L. (1949) Establishment of conditioned responses in chick embryos. *Journal of Comparative and Physiological Psychology* 42, 107–117.

Hunter, A. and Stewart, M.G. (1991) Time-course of synaptic plasticity in the chick LPO following memory formation. *Abstracts of the Avian Learning and Plasticity Conference*. Open University, Milton Keynes, UK, pp. 57.

Hutchison, J.B., Steimer, T.J. and Hutchison, R.E. (1986) Formation of behaviorally active oestrogen in the dove brain: induction of preoptic aromatase by intra-cranial testosterone. *Neuroendocrinology* 43, 416–427.

Hyden, H., Cupello, A. and Palm, A. (1984) Increased binding of GABA to its post-synaptic carrier sites on the plasma membrane of Dieter's neurons after a learning experiment in rats. *Brain Research* 294, 37–45.

Impekoven, M. and Gold, P.S. (1973) Prenatal origins of parent–young interactions in birds: a naturalistic approach. In: Gottlieb, G. (ed.) *Behavioral Embryology*, Vol. 1, Academic Press, New York, pp. 325–356.

Jackson, H. and Rubel, E.W. (1978). Ontogeny of behavioral responsiveness to sound in the chick embryo as indicated by electrical recordings of motility. *Journal of Comparative and Physiological Psychology* 92, 682–692.

Jackson, H., Hackett, J.T. and Rubel, E.W. (1982) Organisation and development of brain stem auditory nuclei in the chick: ontogeny of postsynaptic responses.

Journal of Comparative Neurology 210, 80–86.

Jaynes, J. (1956) Imprinting: the interaction of learned and innate behavior. I. Development and generalization. *Journal of Comparative and Physiological Psychology* 49, 201–206.

Johnson, M.H. and Horn, G. (1986) Dissociation between recognition memory and associative learning by a restricted lesion to the chick forebrain. *Neuropsychologia* 24, 329–340.

Johnson, M.H. and Horn, G. (1988) Development of filial preferences in dark-reared chicks. *Animal Behaviour* 36, 675–683.

Johnson, M.H., Bolhuis, J.J. and Horn, G. (1985) Interaction between acquired preferences and developing predispositions during imprinting. *Animal Behaviour* 33, 1000–1006.

Johnson, M.H., Davies, D.C. and Horn, G. (1989) A sensitive period for the development of a predisposition in dark-reared chicks. *Animal Behaviour* 37, 1044–1058.

Johnson, M.H., Bolhuis, J.J. and Horn, G. (1992) Predispositions and learning: behavioural dissociations in the chick. *Animal Behaviour* 44, 943–948.

Johnston, A.N. and Rogers, L.J. (1992) Glutamate receptor binding and memory formation in the chick. *Proceedings of the Australian Neuroscience Society* 3, 145.

Johnston, A.N., Johnston, G.A.R. and Rogers, L.J. (1993) Glutamate and imprinting memory: The role of glutamate receptors in the encoding of imprinting memory. *Behavioural Brain Research* 54, 137–143.

Johnston, A.N., Bourne, R.C., Stewart, M.G., Rose, S.P.R. and Rogers, L.J. (1994a) Effects of light stimulation on forebrain receptor binding levels in chick embryos. *Brain Research* submitted.

Johnston, A.N., Rogers, L.J. and Dodd, P.R. (1994b) [^{3}H]MK-801 binding asymmetry in the IMHV region of dark-reared chicks is reversed by imprinting. *Brain Research Bulletin*, in press.

Johnston, T.D. and Gottlieb, G. (1985) Effects of social experience on visually imprinted maternal preferences in Peking ducklings. *Developmental Psychobiology* 18, 261–271.

Jones, R.B. (1977) Sex and strain differences in the open-field responses of the domestic chick. *Applied Animal Ethology* 3, 255–261.

Jones, R.B. (1979) The hole-in-the-wall test: its validity as a measure of the 'timidity' aspect of fear in the domestic chick. *UIRCS Medical Science* 7, 167.

Jones, R.B. (1982) Effects of early environmental enrichment upon open-field behaviour and timidity in the domestic chick. *Behavioural Processes* 5, 161–172.

Jones, R.B. (1984) Open-field responses of domestic chicks in the presence of a cage-mate or a strange chick. *IRCS Medical Science* 12, 482–483.

Jones, R.B. (1987a) The assessment of fear in the domestic chick. In: Zayan, R. and Duncan, I.J.H. (eds) *Cognitive Aspects of Social Behaviour in the Domestic Fowl*, Elsevier, Amsterdam, pp. 40–81.

Jones, R.B. (1987b) Social and environmental aspects of fear in the domestic fowl. In: Zayan, R. and Duncan, I.J.H. (eds) *Cognitive Aspects of Social Behaviour in Domestic Fowl*, Elsevier, Amsterdam, pp. 82–149.

Jones, R.B. (1987c) Food neophobia and olfaction in domestic chicks. *Bird Behaviour* 7, 78–81.

Jones, R.B. (1989) Development of alleviation of fear in poultry. In: Faure, J.M. and

Mills, D. (eds) *Proceedings of the Third European Symposium of Poultry Welfare*. The French Branch of the World's Poultry Science Association, Tours, France, pp. 123–136.

Jones, R.B. (1993) Reduction of the domestic chick's fear of human beings by regular handling and related treatments. *Animal Behaviour* 46, 991–998.

Jones, R.B. and Black, A.J. (1979) Behavioral responses of the domestic chick to blood. *Behaviorial and Neural Biology* 27, 319–329.

Jones, R.B. and Black, A.J. (1980) Feeding behaviour of domestic chicks in a novel environment: effects of food deprivation and sex. *Behavioural Processes* 5, 173–183.

Jones, R.B. and Faure, J.M. (1981a) The effects of regular handling on fear responses in the domestic chick. *Behavioural Processes* 6, 135–143.

Jones, R.B. and Faure, J.M. (1981b) Sex effects on open-field behaviour in the domestic chick as a function of age. *Biology and Behavior* 6, 265–272.

Jones, R.B. and Faure, J.M. (1982a) Open-field behaviour of male and female domestic chick as a function of housing conditions, tests situations and novelty. *Biology and Behaviour* 7, 17–25.

Jones, R.B. and Faure, J.M. (1982b) Domestic chicks prefer familiar soiled substrate in an otherwise novel environment. *IRCS Medical Science* 10, 847.

Jones, R.B. and Gentle, M.J. (1985) Olfaction and behavioral modification in domestic chicks (*Gallus domesticus*). *Physiology and Behavior* 34, 917–924.

Jones, R.B. and Hughes, B.O. (1981) Effects of regular handling on growth in male and female chicks of broiler and layer strains. *British Poultry Science* 22, 461–465.

Jones, R.B. and Merry, B.J. (1988) Individual or paired exposure of domestic chicks to an open field: some behavioural and adrenocortical consequences. *Behavioural Processes* 16, 75–86.

Jones, R.B. and Mills, A.D. (1983) Estimation of fear in two lines of the domestic chick: correlations between various methods. *Behavioural Processes* 8, 243–253.

Jones, R.B. and Waddington, D. (1992) Modification of fear in domestic chicks, *Gallus gallus domesticus*, via regular handling and early environmental enrichment. *Animal Behaviour* 43, 1021–1033.

Jones, R.B. and Waddington, D. (1993) Attenuation of the domestic chick's fear of human beings via regular handling: in search of a sensitive period. *Applied Animal Behaviour Science* 36, 185–195.

Junco, F. (1993) Filial imprinting in blackbird nestlings, *Turdus merula*, after only one feeding session. *Animal Behavior* 45, 619–622.

Kahn, A.J. (1973) Ganglion cell formation in the chick retina. *Brain Research* 63, 285–290.

Kahn, A.J. (1974) An autoradiographic analysis of the time of appearance of neurons in the developing chick neural retina. *Developmental Biology* 38, 30–40.

Kalil, R. (1980) A quantitative study of the effects of monocular enucleation and deprivation on cell growth in the dorsal lateral geniculate nucleus of the cat. *Journal of Comparative Neurology* 189, 483–524.

Karakashian, S.J., Gyger, M. and Marler, P. (1988) Audience effects on alarm calling in chickens (*Gallus gallus*). *Journal of Comparative Psychology* 102, 129–135.

Kare, M.R. and Rogers, J.G. Jr (1976) Sense organs. In: Sturkie, P.D. (ed.) *Avian Physiology*. Springer-Verlag, New York, pp. 29–52.

Kare, M.R., Black, R. and Allison, E.G. (1957) The sense of taste in the fowl. *Poultry*

Science 36, 129–138.

Karten, H.J., Hodos, W., Nauta, W.J.H. and Revzin, A.M. (1973) Neural connections of the 'visual Wulst' of the avian telencephalon: experimental studies in the pigeon and owl. *Journal of Comparative Neurology* 150, 253–278.

Kasamatsu, T., Pettigrew, J.D. and Ary, M. (1979) Restoration of visual cortical plasticity by local microperfusion of norepinephrine. *Journal of Comparative Neurology* 185, 163–182.

Kavanagh, J.M., Dodd, P.R. and Rostas, J.A.P. (1991) [³H]MK-801 binding in immature and mature chicken forebrain. *Neuroscience Letters* 134, 83–87.

Kelly, D.B. and Nottebohm, F. (1979) Projections of a telencephalic auditory nucleus – field L – in the canary. *Journal of Comparative Neurology* 183, 455–470.

Kent, J.P. (1987) Experiments on the relationship between the hen and chick (*Gallus gallus*): the role of the auditory mode in recognition and the effects of maternal separation. *Behaviour* 102, 1–14.

Kent, J.P. (1992) The relationship between the hen and the chick, *Gallus gallus domesticus*: the hen's recognition of the chick. *Animal Behaviour* 44, 996–998.

Kerr, L.M., Ostapoff, M. and Rubel, E.W. (1979) Influence of acoustic experience on the ontogeny of frequency generalization gradients in the chicken. *Journal of Experimental Psychology: Animal Behavior Processes* 5, 97–115.

Kilham, P., Klopfer, P.H. and Oelke, H. (1968) Species identification and colour preferences in chicks. *Animal Behaviour* 16, 238–244.

Kimura, D. (1992) Sex differences in the brain. *Scientific American* 267(3), 80–87.

Kimura, N., Kurosawa, N., Kondo, K. and Tsukada, Y. (1993) Molecular cloning of the kainate-binding protein and calmodulin genes which are induced by an imprinting stimulus in ducklings. *Molecular Brain Research* 17, 351–355.

Kirn, J.R., Alvarez-Buylla, A. and Nottebohm, F. (1991) Production and survival of projection neurones in a forebrain vocal center of adult male canaries. *Journal of Neuroscience* 11, 1756–1762.

Klein, R.M. and Andrew, R.J. (1986) Distraction, decisions and persistence in runway tests using the domestic chick. *Behaviour* 99, 139–156.

Kleinschmidt, A., Bear, M. and Singer, W. (1987) Blockade of 'NMDA' receptors disrupts experience-dependent plasticity of kitten striate cortex. *Science* 238, 355–358.

Klopfer, P.H. and Gottlieb, G. (1962) Imprinting and behavioral polymorphism: auditory and visual imprinting in domestic ducks (*Anas platyrhynchos*) and the involvement of the critical period. *Journal of Comparative and Physiological Psychology* 55, 126–130.

Kohsaka, S-I., Takamatsu, K., Aoki, E. and Tsukada, Y. (1979) Metabolic mapping of chick brain after imprinting using [¹⁴C]2-deoxyglucose technique. *Brain Research* 172, 539–544.

Korte, M. and Rauschecker, J.P. (1993) Auditory spatial tuning of cortical neurons is sharpened in cats with early blindness. *Journal of Neurophysiology* 70(4), 1717–1721.

Kossel, A., Egert, U. and Rauschecker, J.P. (1987) Ketamine and not xylazine impairs consolidation of plastic changes in kitten visual cortex. *Society for Neuroscience Abstracts* 13, 1242.

Kovach, J.K. (1968) Spatial orientation of the chick embryo during the last five days of incubation. *Journal of Comparative and Physiological Psychology* 66, 283–288.

Kovach, J.K. (1971) Effectiveness of different colours in the elicitation and development of approach behaviour in chicks. *Behaviour* 38, 154–168.

Kovach, J.K. and Wilson, G. (1993) Early approach preferences of patterns and colours in quail: responses to artificial selection and imprinting. *Animal Behaviour* 46, 95–109.

Krebs, J.R. (1990) Food-storing birds: adaptive specialization in brain and behaviour? *Philosophical Transactions of the Royal Society of London B* 329, 153–160.

Kruijt, J.P. (1964) Ontogeny of social behaviour in Burmese Red Jungle fowl (*Gallus gallus spadiceus*). *Behaviour* (Suppl.) 12, 1–201.

Kruijt, J.P. and Meeuwissen, G. (1991) Sexual preferences of male zebra finches: effects of early and adult experience. *Animal Behaviour* 42, 91–102.

Kuenzel, W.J. (1989) Neuroanatomical substrates involved in the control of food intake. *Poultry Science* 68, 926–937.

Kuo, Z-Y. (1932a) Ontogeny of embryonic behavior in Aves. I. The chronology and general nature of the behavior of the chick embryo. *Journal of Experimental Zoology* 61, 395–430.

Kuo, Z-Y. (1932b) Ontogeny of embryonic behavior in Aves. III. The structural and environmental factors in embryonic behavior. *Journal of Comparative Psychology* 13, 245–272.

Kuo, Z-Y. (1976) *The Dynamics of Behavior Development*, Plenum, New York.

Kuriyama, K., Sisken, B., Ito, J., Simonsen, D.G., Haber, B. and Roberts, E. (1968). The γ-aminobutyric acid system in the developing chick embryo cerebellum. *Brain Research* 11, 412–430.

Lauber, J.K. and Shutze, J.V. (1964) Accelerated growth of embryo chicks under the influence of light. *Growth* 28, 179–190.

Lehrman, D.S. (1953) A critique of Konrad Lorenz's theory of instinctive behavior. *Quarterly Review of Biology* 28, 337–363.

Lemon, R.E. (1973) Nervous control of the syrinx of white-throated sparrows (*Zonotrichia albicollis*). *Journal of Zoology* 171, 131–140.

Levi, G. and Morisi, G. (1971) Free amino acids and related compounds in chick brain during development. *Brain Research* 26, 131–140.

Lickliter, R. (1990) Premature visual stimulation accelerates intersensory functioning in bobwhite quail neonates. *Developmental Psychobiology* 23, 15–27.

Lickliter, R. (1993) Timing and the development of perinatal perceptual organisation. In: Turkewitz, G. and Devenny, D.A. (eds) *Developmental Time and Timing*. L. Erlbaum, Hillsdale, NJ, pp. 105–123.

Lickliter, R. and Gottlieb, G. (1985) Social interaction with siblings is necessary for visual imprinting of species-specific maternal preferences in ducklings (*Anas platyrhynchos*). *Journal of Comparative Psychology* 99, 371–379.

Lickliter, R. and Gottlieb, G. (1986) Training ducklings in broods interferes with maternal imprinting. *Developmental Psychobiology* 19, 555–566.

Lickliter, R. and Gottlieb, G. (1987) Retroactive excitation: post-training social experience with siblings consolidates maternal imprinting in ducklings. *Journal of Comparative Psychology* 101, 40–46.

Lickliter, R. and Gottlieb, G. (1988) Social specificity: interaction with one species is necessary to foster species-specific maternal preference in ducklings. *Developmental Psychobiology* 21, 311–321.

Lickliter, R. and Stoumbos, J. (1991) Enhanced prenatal auditory experience facilities species-specific visual responsiveness in bobwhite quail chicks (*Colinus virginianus*). *Journal of Comparative Psychology* 105, 89–94.

Lickliter, R. and Virkar, P. (1989) Intersensory functioning in bobwhite quail chicks: early sensory dominance. *Developmental Psychobiology* 22, 641 667.

Lickliter, R., Dyer, A.B. and McBride, T. (1993) Perceptual consequences of early social experience in precocial birds. *Behavioural Processes* 30, 185–200.

Lindenmaier, P. and Kare, M. (1959) The taste end-organs of the chicken. *Poultry Science* 38, 545–550.

Linser, P.J. and Perkins, M. (1987) Gliogenesis in the embryonic avian optic tectum: neuronal–glial interactions influence astroglial phenotype maturation. *Developmental Brain Research* 31, 277–290.

Lombardi, C.M., Fachinelli, C.C. and Delius, J.D. (1984) Odditiy of visual patterns conceptualized by pigeons. *Animal Learning and Behavior* 12, 2–6.

Longstaff, A. and Rose, S.P.R. (1981) Ontogenetic- and imprinting-induced changes in chick brain protein metabolism and muscarinic receptor binding activity. *Journal of Neurochemistry* 37, 1089–1098.

Lorenz, K. (1935) Der Kumpan in der Umwelt des Vogels. *Journal für Ornithologie* 83, 137–213, 289–413.

Lowndes, M., Stewart, M.G. and Stanford, D. (1991) Morphological plasticity after passive avoidance learning. *Abstracts of the Avian Learning and Plasticity Conference*, Open University, Milton Keynes, UK, pp. 59.

MacKintosh, N.J. (1983) *Conditioning and Associative Learning.* Oxford University Press, Oxford.

Macko, K.A. and Hodos, W. (1984) Near-field acuity after visual system lesions in pigeons: I. Thalamus. *Behavioural Brain Research* 13, 1–14.

MacNeilage, P.F., Studdert-Kennedy, M.G. and Lindblom, B. (1987) Primate handedness: a foot in the door. *Behavioural Brain Sciences* 11, 737–746.

Macphail, E.M. and Reilly, S. (1989) Medial versus lateral hyperstriatal lesions in pigeons: effects on autoshaping, non-matching-to-sample and spatial discrimination learning at short and long intertrial intervals. *Behavioural Brain Research* 35, 63–73.

Maier, V. and Scheich, H. (1987) Acoustic imprinting in guinea fowl chicks: age dependence of 2-deoxyglucose uptake in relevant forebrain areas. *Developmental Brain Research* 31, 15–27.

Margoliash, D. and Fortune, E.S. (1992) Temporal and harmonic combination – sensitive neurons in the zebra finch's HVc. *Journal of Neuroscience* 12, 4309–4326.

Margoliash, D. and Konishi, M. (1985) Auditory representation of autogenous song in the song system of white-crowned sparrows. *Proceedings of the National Academy of Science, USA* 82, 5997–6000.

Markus, E.J., Petit, T.L. and LeBoutillier, J.B. (1987) Synaptic structural changes during development and ageing. *Developmental Brain Research* 35, 239–248.

Marler, P., Dufty, A. and Pickert, R. (1986) Vocal communication in the domestic chicken: II. Is a sender sensitive to the presence and nature of a receiver? *Animal Behaviour* 34, 194–198.

Marler, P., Karakashian, S. and Gyger, M. (1991) Do animals have the option of withholding signals when communication is inappropriate? The audience effect. In: Ristau, C.A. (ed.) *Cognitive Ethology.* Erlbaum, Hillsdale, NJ, pp. 187–208.

Martin, G.R. (1985) Eye. In: King, A.S. and McLelland, T. (eds) *Form and Function in Birds*, Volume 3. Academic Press, London, pp. 311–373.

Martin, G.R. (1993) Producing the image. In: Zeigler, H.P. and Bischof, H.-J. (eds) *Vision, Brain, and Behaviour in Birds*. MIT Press, Cambridge, MA, pp. 5–24.

Martin, J.T. (1978) Embryonic pituitary adrenal axis, behavior development and domestication in birds. *American Zoologist* 18, 489–499.

Mason, G.J. (1991) Sterotypies: a critical review. *Animal Behaviour* 41, 1015–1037.

Mason, R.J. and Rose, S.P.R. (1987) Lasting changes in spontaneous multi-unit activity in the chick brain following avoidance training. *Neuroscience* 21, 931–941.

Maxwell, J.H. and Granda, A.M. (1979) Receptive fields of movement – sensitive cells in the pigeon thalamus. In: Granda, A.M. and Maxwell, J.H. (eds) *Neural Mechanisms of Behavior in the Pigeon*. Plenum Press, New York, pp. 177–197.

McBride, G., Parer, I.P. and Foenander, F. (1969) The social organisation of behaviour of the feral domestic fowl. *Animal Behaviour Monograph* 2, 127–181.

McBride, G., Foenander, F. and Slee, C. (1970) The development of social and sexual behaviour in the domestic fowl. *Revue de Comportment Animal* 4, 51–57.

McBride, T.C. and Lickliter, R. (1993) Social experience with siblings fosters species-specific responsiveness to maternal visual cues in bobwhite quail chicks (*Colinus virginianus*). *Journal of Comparative Psychology* 107, 320–327.

McCabe, B.J. and Horn, G. (1988) Learning and memory: regional changes in N-methyl-D-aspartate receptors in the chick brain after imprinting. *Proceedings of the National Academy of the Sciences USA* 85, 2849–2853.

McCabe, B.J. and Horn, G. (1991) Synatic transmission and recognition memory: time course of changes in N-methyl-D-aspartate receptors after imprinting. *Behavioral Neuroscience* 105, 289–294.

McCabe, B.J. and Horn, G. (1993) Imprinting leads to elevated Fos-like immuno-reactivity in the intermediate and medial hyperstriatum ventrale (IMHV) of the domestic chick. *Journal of Physiology* 459, 160P.

McCabe, B.J., Davey, J.E. and Horn, G. (1992) Impairment of learning by localized injection of an N-methyl-D-aspartate receptor antagonist into the hyperstriatum ventrale of the domestic chick. *Behavioral Neuroscience* 106, 947–953.

McCabe, N.R. and Rose, S.P.R. (1985) Passive avoidance training increases fucose incorporation into glycoproteins in chick forebrain slices in vitro. *Neurochemistry Research* 10, 1083–1095.

McCasland, J.S. and Konishi, M. (1981) Interaction between auditory and motor activities in an avian song control nucleus. *Proceedings of the National Academy of Science, USA* 78, 7815–7819.

McCoshen, J.A. and Thompson, R.P. (1968) A study of clicking and its source in some avian species. *Canadian Journal of Zoology* 46, 169–172.

McFarland, D. (1989) *Problems of Animal Behaviour*, Longman Scientific and Technical, New York, pp. 87–104.

McGraw, C.F. and McLaughlin, B.J. (1980) Fine structural studies of synaptogenesis in the superficial layers of the chick optic tectum. *Journal of Neurocytology* 9, 79–93.

McNaughton, B.L., Barnes, C.A., Rao, G., Baldwin, J. and Rasmussen, M. (1986) Long-term enhancement of hippocampal synaptic transmission and the acquisition of spatial information. *Journal of Neuroscience* 6, 563–571.

McQuoid, L.M. and Galef, B.G. Jr (1993) Social stimuli influencing feeding behaviour of Burmese fowl: a video analysis. *Animal Behaviour* 46, 13–22.

Melan C., de Barry, J. and Ungere, A. (1991) α-L-glutamyl-L-aspartate interacting with NMDA receptors affects appetitive visual discrimination tasks in mice. *Behavioral and Neural Biology* 55, 356–365.

Melsbach, G., Hahmann, U., Waldmann, C., Wörtwein, G. and Güntürkün, O. (1991) Morphological asymmetries of the optic tectum of the pigeon. In: Elsner, N. and Penzlin, H. (eds) *Synapse – Transmission – Modulation*. Thiem, Stuttgart, p. 553.

Mench, J.A. and Andrew, R.J. (1986) Lateralization of a food search task in the domestic chick. *Behavioral and Neural Biology* 46, 107–114.

Mendoza, A.S., Breipohl, W. and Miragall, F. (1982) Cell migration from the chick olfactory placode: a light and electron microscopic study. *Journal of Embryology and Experimental Morphology* 69, 47–59.

Mey, J. and Thanos, S. (1993) Developmental anatomy of the chick retinotectal projection. In: Zeigler, H.P. and Bischof, H.-J. (eds) *Vision, Brain, and Behaviour in Birds*. MIT Press, Cambridge, MA, pp. 173–194.

Micelli, D., Repérant, J., Villalobos, J. and Dionne, L. (1987) Extratelecephalic projections of the avian visual Wulst. A quantitative autoradiographic study on the pigeon, *Columbia livia*. *Journal für Hirnforschung* 28, 45–58.

Miles, F.A. (1972) Centrifugal control of the avian retina. III. Effects of electrical stimulation of the isthmo–optic tract on the receptive field properties of retinal ganglion cells. *Brain Research* 48, 115–129.

Miller, D.B. and Blaich, C.F. (1988) Alarm call responsivity of mallard ducklings: VII. Auditory experience maintains freezing. *Developmental Psychobiology* 21, 523–533.

Miller, D.B., Hicinbothom, G. and Blaich, C.F. (1990) Alarm call responsivity of mallard ducklings: multiple pathways in behavioural development. *Animal Behaviour* 39, 1207–1212.

Millikan, G.C. and Bowman, R.I. (1967) Observations on Galapagos tool-using finches in captivity. *Living Birds* 6, 23–41.

Miyake, M. and Morino, H. (1992) Developmental changes in β-citryl-L-glutamate concentration and its synthetic and hydrolytic activities in neuronal cells cultured from chick embryo optic lobes. *Journal of Neurochemistry* 59, 1654–1660.

Moltz, H. and Stettner, L.J. (1961) The influence of patterned light deprivation on the critical period for imprinting. *Journal of Comparative and Physiological Psychology* 54, 279–283.

Monaghan, D.T., Bridges, R.J. and Cotman, C.W. (1989) The excitatory amino acid receptors: their classes, pharmacology and distinct properties in the function of the central nervous system. *Annual Reviews in Pharmacology and Toxicology* 29, 365–402.

Morris, R.G.M., Davis, S. and Butcher, S.P. (1990) Hippocampal synaptic plasticity and MNDA receptors: a role in information storage? *Philosophical Transactions of the Royal Society of London B* 329, 89–106.

Morris, R.G.M., Garrud, P., Rawlins, J. and O'Keefe, J. (1982) Place navigation impaired in rats with hippocampal lesions. *Nature* 297, 681–683.

Morris, V.B. (1987) An afoveate area centralis in the chick retina. *Journal of Comparative Neurology* 210, 198–203.

Mukhametov, L.M. (1987) Unihemispheric slow-wave sleep in the Amazonian dolphin, *Inia goeffrensis*. *Neuroscience Letters* 79, 128–132.

Müller, C.M. and Scheich, H. (1988) Contribution of GABAergic inhibition to the response characteristics of auditory units in the avian forebrain. *Journal of Neurophysiology* 59, 1673–1689.

Myhre, K. (1978) Behavioral temperature regulation in neonate chick of bantam hen (*Gallus domesticus*). *Poultry Science* 57, 1369–1375.

Naftolin, F., Leranth, C., Perez, J. and Garcia-Segura, L.M. (1993) Estrogen induces synaptic plasticity in adult primate neurons. *Neuroendocrinology* 57, 935–939.

Nakazawa, T., Tachi, S., Aikawa, E. and Ihnuma, M. (1993) Formation of the myelinated nerve fibre layer in the chicken retina. *Glia* 8, 114–121.

Narang, H.K. (1977) Right–left asymmetry of myelin development in epiretinal portion of rabbit optic nerve. *Nature*, 266, 855–856.

Narang, H.K. and Wisniewski, H.M. (1977) The sequence of myelination in the epiretinal portion of the optic nerve in the rabbit. *Neuropathology and Applied Neurobiology* 3, 15–27.

Nauta, W.J.H. and Karten, H.J. (1970) A general profile of the vertebrate brain with sidelights of an ancestry of the cerebral cortex. In: Schmidtt, O. (ed.) *The Neuroscience Second Study Program*, Rockefeller University Press, New York, pp. 7–26.

Ng, K.T. and Gibbs, M.E. (1991) Stages in memory formation: a review. In: Andrew, R.J. (ed.) *Neural and Behavioural Plasticity: The Use of the Domestic Chick as a Model*, Oxford University Press, Oxford, pp. 351–369.

Ng, K.T., Gibbs, M.E., Crowe, S.F., Sedman, G.L., Hua, F., Zhao, W., O'Dowd, B., Richard, N., Gibbs, C.L., Syková, E., Svoboda, J. and Jendelová, P. (1992) Molecular mechanisms of memory formation. *Molecular Neurobiology* 5, 333–350.

Nicol, C.J. and Pope, S.J. (1993) Food deprivation during observation reduces social learning in hens. *Animal Behaviour* 45, 193–196.

Nixdorf, B. and Bischof, H-J. (1986) Posthatching development of synapses in the neuropil of nucleus rotundus of the zebra finch: a quantitative electron micro-scope study. *Journal of Comparative Neurology* 250, 133–139.

Noller, N.J. (1984) Eye enlargement in the chick following visual deprivation. MSc. dissertation, University of Melbourne, Australia.

Nordeen, E.J. and Nordeen, K.W. (1990) Neurogenesis and sensitive periods in avian song learning. *Trends in Neuroscience* 13, 31–36.

Nottebohm, F. (1970) Ontogeny of bird song. *Science* 167, 950–956.

Nottebohm, F. (1971) Neural lateralization of vocal control in a Passerine bird. I. Song. *Journal of Experimental Zoology* 177, 229–261.

Nottebohm, F. (1972) Neural lateralization of vocal control in a Passerine bird. II. Subsong, calls and a theory of vocal learning. *Journal of Experimental Zoology* 179, 35–50.

Nottebohm, F. (1976) Phonation in the orange-winged amazon parrot, *Amazona amazonica*. *Journal of Comparative Physiology* 108, 157–170.

Nottebohm, F. (1977) Asymmetries in neural control of vocalization in the canary. In: Harnard, S., Doty, R.W., Goldstein, L., Jaynes, J. and Krauthamer, G. (eds) *Lateralisation in the Nervous System*, Academic Press, New York, pp. 23–44.

Nottebohm, F. (1987) Plasticity in adult avian central nervous system: possible relation between hormones, learning, and brain repair. In: Mountcastle, V.B.,

Plum, F. and Geiger, S.R. (eds) *Handbook of Physiology V; Section 1. The Nervous System*, American Physiological Society, Maryland, pp. 85–108.

Nottebohm, F. (1989) From bird song to neurogenesis. *Scientific American* Feb., 55–61.

Nottebohm, F. and Nottebohm, M.E. (1976) Left hypoglossal dominance in the control of canary and white-crowned sparrow song. *Journal of Comparative Physiology* 108, 171–192.

Nottebohm, F. and Nottebohm, M.E. (1978) Relationship between song repertoir and age in the canary, *Serinus canarius*. *Zietschrift für Tierpsychologie* 46, 298–305.

Nottebohm, F., Stokes, T.M. and Leonard, C.M. (1976) Central control of song in the canary, *Serinus canarius*. *Journal of Comparative Neurology* 165, 457–486.

Nottebohm, F., Manning, E. and Nottebohm, M.E. (1979) Reversal of hypoglossal dominance in canaries following unilateral syringeal denervation. *Comparative Physiology* 134, 227–240.

Nottebohm, F., Alvarez-Buylla, A., Cynx, J., Kirn, J., Ling, C.Y. and Nottebohm, M. (1990) Song learning in birds: the relation between perceptions and production. *Philosophical Transactions of the Royal Society of London B* 329, 115–124.

Nowicki, S. and Capranica, R.R. (1986) Bilateral syringeal interaction in vocal production of an oscine bird sound. *Science* 231, 1297–1299.

Nye, P.W. (1973) On the functional differences between frontal and lateral visual fields of the pigeon. *Vision Research* 13, 559–574.

O'Connor, R.J. (1984) *The Growth and Development of Birds*. Wiley, Chichester.

O'Donovan, M.J. and Landmesser, L. (1987) The development of hindlimb motor activity studies in the isolated spinal cord of the chick embryo. *Journal of Neuroscience* 7, 3256–3264.

Okado, N. and Oppenheim, R.W. (1985) The onset and development of descending pathways to the spinal cord in the chick embryo. *Journal of Comparative Neurology* 232, 143–161.

Okano, M. and Kasuga, S. (1980) Developmental studies on the olfactory epithelium in domestic fowls. In: Tanake, Y., Tanaka, K. and Ookawa, T. (eds) *Biological Rhythms in Birds: Neural and Endocrine Aspects*, Springer-Verlag, Berlin, pp. 235–248.

O'Leary, D.M., Gerten, C.R. and Cowan, C.M. (1983) The development and restriction of the ipsilateral retinotectal projection in the chick. *Developmental Brain Research* 10, 93–109.

Oleksenko, A.I., Mukhametov, L.M., Polyokova, I.G., Supin, A.Y. and Kovalzon, V.M. (1992) Unihemispheric sleep deprivation in bottlenose dolphins. *Journal of Sleep Research* 1, 40–44.

Olney, J.W. (1978) Neurotoxicity of excitatory amino acids. In: Olney, J.W. and McGeer, P.L. (eds) *Kainic Acid as a Tool in Neurobiology*. Raven Press, New York, pp. 95–122.

Ookawa, T. (1971) Electroencephalograms recorded from the telencephalon of the blinded chicken during behavioral sleep and wakefulness. *Poultry Science* 50, 731–736.

Oppenheim, R. (1966) Amniotic contraction and embryonic motility in the chick embryo. *Science* 152, 528–529.

Oppenheim, R.W. (1968) Light responsivity in chick and duck embryos just prior to hatching. *Animal Behaviour* 16, 276–280.

Oppenheim, R.W. (1970) Some aspects of embryonic behaviour in the duck (*Anas platyrhynchos*). *Animal Behaviour* 18, 335–352.

Oppenheim, R.W. (1972a) Experimental studies of hatching behavior in the chick. III. The role of the midbrain and forebrain. *Journal of Comparative Neurology* 146, 479–505.

Oppenheim, R.W. (1972b) Prehatching and hatching behaviour in birds: a comparative study of altricial and precocial species. *Animal Behaviour* 20, 644–655.

Oppenheim, R.W. (1973) Prehatching and hatching behavior: a comparative and physiological consideration. In: Gottlieb, G. (ed.) *Behavioral Embryology*, Volume 1. Academic Press, New York, pp. 163–244.

Oppenheim, R.W., Schwartz, L.M. and Shatz, C.J. (1992) Neuronal death, a tradition of dying. *Journal of Neurobiology* 23, 1111–1115.

Panzica, G.C. and Viglietti-Panzica, C.V. (1981) Electron microscopy of synaptic structures in optic tectum of developing chick embryos. *Bibliotheca Anatomica* 19, 167–173.

Paredes, R.G. and Ågmo, A. (1992) GABA and behaviour: The role of receptor subtypes. *Neuroscience and Biobehavioural Review* 16, 145–170.

Parsons, C.H. and Rogers, L.J. (1992) Ketamine/xylazine extends the sensitive period for imprinting memory in the chick. *Proceedings of the Australian Society of Neuroscience* 3, 82.

Parsons, C.H. and Rogers, L.J. (1993) Role of the tectal and posterior commissures in lateralization of the avian brain. *Behavioural Brain Research* 54, 153–164.

Patel, S.N. and Stewart, M.G. (1988) Changes in the number and structure of dendritic spines 25h after passive avoidance training in the domestic chick, *Gallus domesticus*. *Brain Research* 449, 34–46.

Patetta-Queriolo, M.A. and Garcia-Austt, E. (1956) Development of the electroretinogram in the chick embryo. *EGG Clinical Neurophysiology* 8, 155.

Patterson, T.A. and Rose, S.P.R. (1992) Memory in the chick: multiple cues, distinct brain locations. *Behavioural Neuroscience* 106, 465–470.

Patterson, T.A., Alvarado, M.C., Warner, I.T., Bennett, E.L. and Rosenzweig, M.R. (1986) Memory stages and brain asymmetry in chick learning. *Behavioral Neuroscience* 100, 856–865.

Patterson, T.A., Gilbert, D.B. and Rose, S.P.R. (1990) Pre- and post-training lesions of the intermediate medial hyperstriatum ventrale and passive avoidance learning in the chick. *Experimental Brain Research* 80, 189–195.

Pearson, R. (1972) *The Avian Brain*. Academic Press, London.

Pepperberg, I.M. (1990a) Conceptual abilities of some non-primate species, with an emphasis on an African Grey parrot. In: Parker, S.T. and Gibson, K.R. (eds) *'Language' and Intelligence in Monkeys and Apes*. Cambridge University Press, Cambridge, pp. 469–507.

Pepperberg, I.M. (1990b) Some cognitive capacities of an African grey parrot (*Psittacus erithacus*). *Advances in the Study of Behaviour* 19, 357–409.

Pepperberg, I.M. (1990c) Cognition in an African Gray Parrot (*Psittacus erithacus*): further evidence for comprehension of categories and labels. *Journal of Comparative Psychology* 104(1), 41–52.

Péquignot, Y. and Clarke, P.G.H. (1992a) Changes in lamination and neuronal survival in the isthmo-optic nucleus following the intraocular injection of tetrodotoxin in chick embryos. *Journal of Comparative Neurology* 321, 336–350.

Péquignot, Y. and Clarke, P.G.H. (1992b) Maintenance of targeting errors by isthmo-optic axons following the intraocular injection of tetrodotoxin in chick embryos. *Journal of Comparative Neurology* 321, 351–356.

Perera, A.D. and Follett, B.K. (1992) Photoperiodic induction *in vitro*: the dynamics of gonadotropin-releasing hormone release from hypothalamic explants of the Japanese quail. *Endocrinology* 131, 2898–2908.

Peters, J.J., Vonderahe, A.R. and Huesman, A.A. (1960) Chronological development of electrical activities in the optic lobes, cerebellum, and cerebrum of the chick embryo. *Physiological Zoology* 33, 225–231.

Peters, J.J., Vonderahe, A.R. and Powers, T.H. (1958) Electrical studies of functional development of the eye and optic lobes in the chick embryo. *Journal of Experimental Zoology* 139, 459–468.

Peters, J.J., Vonderahe, A.R. and Schmid, D. (1965) Onset of cerebral electrical activity associated with behavioural sleep and attention in the developing chick. *Journal of Experimental Zoology* 160, 255–262.

Petersen, M.R., Beecher, M.D., Zoloth, S.R., Green, S., Marler, P.R., Moody, D.B. and Stebbins, W.C. (1984) Neural lateralization of vocalizations by Japanese macaques: communicative significance is more important than acoustic structure. *Behavioral Neuroscience* 98, 779–790.

Pettigrew, J.D. (1979) Binocular visual processing in the owl's telencephalon. *Proceedings of the Royal Society London Series B* 204, 435–454.

Pettigrew, J.D. and Konishi, M. (1976a) Neurons selective for orientation and binocular disparity in the visual Wulst of the barn owl (*Tyto alba*). *Science* 193, 675–678.

Pettigrew, J.D. and Konishi, M. (1976b) Effects of monocular deprivation on binocular neurons in the owl's visual Wulst. *Nature* 264, 753–754.

Phillips, R.E. (1966) Evoked potential study of the connections of the avian archistriatum and neostriatum. *Journal of Comparative Neurology* 127, 89–100.

Phillips, R.E. and Siegel, P.B. (1966) Development of fear in chicks of two closely related genetic lines. *Animal Behaviour* 14, 84–88.

Phillips, R.E. and Youngren, O.M. (1986) Unilateral kainic acid lesions reveal dominance of right archistriatum in avian fear behavior. *Brain Research* 377, 216–220.

Prada, C., Puga, J., Pérez–Méndez, L., López, R. and Ramirez, G. (1991) spatial and temporal patterns of neurogenesis in the chick retina. *European Journal of Neuroscience* 3, 559–569.

Premack, D. (1978) On the abstractness of human concepts: why it would be difficult to talk to a pigeon. In: Hulse, S.H., Fowler, H. and Haug, W.K. (eds) *Cognitive Processes in Animal Behavior*. Erlbaum, Hillsdale, NJ, pp. 421–451.

Pritz, M.B., Mead, W.R. and Northcutt, R.G. (1970) The effects of Wulst ablations on color, brightness and pattern discrimination in pigeon *(Columba livia)*. *Journal of Comparative Neurology* 140, 81–100.

Provine, R.R. (1972) Ontogeny of bioelectric activity in the spinal cord of the chick embryo and its behavioural implications. *Brain Research* 41, 365–378.

Provine, R.R. (1973) Neurophysiological aspects of behavior development in the chick embryo. In: Gottlieb, G. (ed.) *Behavioral Embryology*, Volume 1. Academic Press, New York, pp. 77–102.

Rager, G. (1976) Morphogenesis and physiogenesis of the retino-tectal connection in the chicken. II. The retino-tectal synapses. *Proceedings of the Royal Society of London, B* 192, 353–370.

Rager, G. (1980) Development of the retinotectal projection in the chicken. *Advances in Anatomy, Embryology and Cell Biology* 63, 11–30, 663–681.

Rager, G. and von Oeynhausen, B. (1979) Ingrowth and ramification of retinal fibers in the developing optic tectum of the chick embryo. *Experimental Brain Research* 35, 213–227.

Rajecki, D.W. (1974) Effects of prenatal exposure to auditory and visual stimuli on social responses in chickens. *Behavioral Biology* 11, 525–536.

Rajecki, D.W. (1988) Formation of leap orders in pairs of male domestic chickens. *Aggressive Behavior* 14, 425–436.

Rajecki, D.W. (1991) Leap orders in male chickens (*Gallus gallus domesticus*): longitudinal evidence. *Journal of Comparative Psychology* 105, 73–77.

Rajendra, S. and Rogers, L.J. (1993) Asymmetry is present in the thalamofugal visual projections of female chicks. *Experimental Brain Research* 92, 542–544.

Rakic, P. (1985) Limits of neurogenesis in primates. *Science* 227, 1054–1056.

Ramsay, A.O. (1951) Familial recognition in domestic birds. *Auk* 68, 1–17.

Ramsay, A.O. and Hess, E.H. (1954) A laboratory approach to the study of imprinting. *Wilson Bulletin* 66, 196–206.

Rashid, N. and Andrew, R.J. (1989) Right hemisphere advantage for topographical orientation in the domestic chick. *Neuropsychologia* 27, 937–948.

Rausch, G. and Scheich, H. (1982) Dendritic spine loss and enlargement during maturation of the speech control system in the Mynah bird (*Gracula religiosa*). *Neuroscience Letters* 29, 129–133.

Rauschecker, J.P. (1991) Mechanisms of visual plasticity: Hebb synapses, NMDA receptors, and beyond. *Physiological Reviews* 71, 587–615.

Rauschecker, J.P. and Hahn, S. (1987) Ketamine–xylazine anaesthesia blocks consolidation of ocular dominance changes in kitten visual cortex. *Nature* 326, 183–185.

Rauschecker, J.P., Egert, U. and Kossel, A. (1990) Effects of NMDA antagonists on developmental plasticity in kitten visual cortex. *International Journal of Developmental Neuroscience* 8, 425–435.

Rauschecker, J.P., Tian, B., Korte, M. and Egert, U. (1992) Crossmodal changes in the somatosensory vibrissa/barrel system of visually deprived animals. *Proceedings of the National Academy of the Sciences, USA* 89, 5063–5067.

Rebillard, G. and Rubel, E.W. (1981) Electrophysiological study of the maturation of auditory responses from the inner ear of the chick. *Brain Research* 229, 15–23.

Regolin, L., Vallortigara, G. and Zanforlin, M. (1994) Perceptual and motivational aspects of detour behaviour in young chicks. *Animal Behaviour* 47, 123–131.

Rehkämper, G., Haase, E. and Frahm, H. (1988) Allometric comparison of brain weight and brain structure volumes in different breeds of the domestic pigeon, *Columba livia* f.d. (fantails, homing pigeons, strassers). *Brain, Behaviour and Evolution* 31, 141–149.

Reiner, A. and Karten, H.J. (1983) The laminar source of efferent projections from the avian visual Wulst. *Brain Research* 275, 349–354.

Reiner, A. and Karten, H.J. (1985) Comparison of olfactory bulb projections in pigeons and turtles. *Brain Behaviour and Evolution* 27, 11–27.

Reymond, E. and Rogers, L.J. (1981a) Diurnal variations in learning performance in chicks. *Animal Behaviour* 29, 241–248.

Reymond, E. and Rogers, L.J. (1981b) Deprivation of the visual and tactile aspects of food important to learning performance of an appetitive task by chicks. *Behavioral and Neural Biology* 31, 425–434.

Reynolds, R.J., McFadden, S.A., Heath, J.W. and Rostas, J.A.P. (1992) Regional development of catecholaminergic innervation in the chick forebrain: a fluorescence histochemical study. *Proceedings of the Australian Neuroscience Society* 3, 136.

Ribatti, D., Nico, B. and Bertossi, M. (1993) The development of the blood-brain barrier in the chick. Studies with Evans blue and horseradish peroxidase. *Annals in Anatomy* 175, 85–88.

Rios, H., Flores, V. and Fiszer de Plazas, S. (1987) Effect of light- and dark-rearing on the postnatal development of GABA receptor sites in the chick optic lobe. *International Journal of Developmental Neuroscience* 5, 319–325.

Robert, F. and Cuenod, M. (1969) Electrophysiology of the intertectal commissures in the pigeon. I. Analysis of pathways. *Experimental Brain Research* 9, 116–122.

Roberts, E. and Kuriyama, K. (1968) Biochemical-physiological correlations in studies of the γ-aminobutyric acid system. *Brain Research* 8, 1–35.

Roberts, S. (1987) Less-than-expected variability in evidence for three stages in memory formation. *Behavioral Neuroscience* 101, 120–125.

Robinzon, B., Snapir, N. and Perek, M. (1977) Removal of olfactory bulbs in chickens: consequent changes in food intake and thyroid activity. *Brain Research Bulletin* 2, 263–271.

Rogers, L.J. (1971) Testosterone, isthmo-optic lesions and visual search in chickens. DPhil thesis, University of Sussex.

Rogers, L.J. (1974) Persistence and search influenced by natural levels of androgens in young and adult chickens. *Physiology and Behavior* 12, 197–204.

Rogers, L.J. (1980) Functional lateralisation in the chicken fore-brain revealed by cycloheximide treatment. In: Nohring, R. (ed.) *Acta XVII Congressus Ornithologici, Berlin 1978*, Vol. 1. Deutsche Ornithologen-Gesellschaft, Berlin, pp. 653–659.

Rogers, L.J. (1981) Environmental influences on brain lateralization. *Behavioural and Brain Sciences* 4, 35–36.

Rogers, L.J. (1982a) Teratological effects of glutamate on behaviour. *Food Technology of Australia* 34, 202–206.

Rogers, L.J. (1982b) Light experience and asymmetry of brain function in chickens. *Nature* 297, 223–225.

Rogers, L.J. (1986) Lateralization of learning in chicks. *Advances in the Study of Behaviour* 16, 147–189.

Rogers, L.J. (1988) Biology, the popular weapon: sex differences in cognitive function. In: Caine, B., Gross, E.A. and de Lepervanche, M. (eds) *Crossing Boundaries: Feminisms and the Critique of Knowledges*, Allen and Unwin, Sydney.

Rogers, L.J. (1989) Laterality in animals. *International Journal of Comparative Psychology* 3, 5–25.

Rogers, L.J. (1990) Light input and the reversal of functional lateralization in the chicken brain. *Behavioural Brain Research* 38, 211–221.

Rogers, L.J. (1991) Development of lateralization. In: R.J. Andrew (ed.) *Neural and Behavioural Plasticity: The Use of the Domestic Chick as a Model*. Oxford University Press, Oxford, pp. 507–535.

Rogers, L.J. (1993) The molecular neurobiology of early learning, development, and sensitive periods, with emphasis on the avian brain. *Molecular Neurobiology 7*, 161–187.

Rogers, L.J. and Adret, P. (1993) Developmental mechanisms of lateralization. In: Zeigler, H.P. and Bischof, H-J. (eds) *Vision, Brain, and Behavior in Birds*. MIT Press, Cambridge, MA, pp. 227–242.

Rogers, L.J. and Andrew, R.J. (1989) Frontal and lateral visual field use by chicks after treatment with testosterone. *Animal Behaviour 38*, 394–405.

Rogers, L.J. and Anson, J.M. (1979) Lateralization of function in the chicken forebrain. *Pharmacology, Biochemistry and Behavior 10*, 679–686.

Rogers, L.J. and Astiningsih, K. (1991) Social hierarchies in very young chicks. *British Poultry Science 32*, 47–56.

Rogers, L.J. and Bell, G.A. (1989) Different rates of functional development in the two visual systems of the chicken revealed by [^{14}C]2-deoxyglucose. *Developmental Brain Research 49*, 161–172.

Rogers, L.J. and Bell, G.A. (1994) Changes in metabolic activity in the hyperstriatum of the chick before and after hatching. *International Journal of Developmental Neuroscience 12(2)*, 557–566.

Rogers, L.J. and Bolden, S.W. (1991) Light-dependent development and asymmetry of visual projections. *Neuroscience Letters 121*, 63–67.

Rogers, L.J. and Chaffey, G. (1994) Lateralised patterns of sleep activity in the developing chick brain. *Proceedings of the Australian Neuroscience Society 5*, 95.

Rogers, L.J. and Drennen, H.D. (1978) Cycloheximide interacts with visual input to produce permanent slowing of visual learning in chickens. *Brain Research 158*, 479–482.

Rogers, L.J. and Ehrlich, D. (1983) Asymmetry in the chicken forebrain during development and a possible involvement of the supraoptic decussation. *Neuroscience Letters 37*, 123–127.

Rogers, L.J. and Hambley, J.W. (1982) Specific and non-specific effects of neuro-excitatory amino acids on learning and other behaviours in the chicken. *Behavioural Brain Research 4*, 1–18.

Rogers, L.J. and Miles, F.A. (1972) Centrifugal control of the avian retina. V. Effects of lesions of the isthmo-optic nucleus on visual behaviour. *Brain Research 48*, 147–156.

Rogers, L.J. and Rajendra, S. (1993) Modulation of the development of light-initiated asymmetry in chick thalamofugal visual projections by oestradiol. *Experimental Brain Research 93*, 89–94.

Rogers, L.J. and Sink, H.S. (1988) Transient asymmetry in the projections of the rostral thalamus to the visual hyperstriatum of the chicken, and reversal of its direction by light exposure. *Experimental Brain Research 70*, 378–384.

Rogers, L.J. and Workman, L. (1989) Light exposure during incubation affects competitive behaviour in domestic chicks. *Applied Animal Behaviour Science 23*, 187–198.

Rogers, L.J. and Workman, L. (1993) Footedness in birds. *Animal Behaviour 45*, 409–411.

Rogers, L.J., Drennen, H. and Mark, R.F. (1974) Inhibition of memory formation in the imprinting period: irreversible action of cycloheximide in young chickens. *Brain Research 79*, 213–253.

Rogers, L.J., Oettinger, R., Szer, J. and Mark, R.F. (1977) Separate chemical inhibitors of long-term and short-term memory: contrasting effects of cycloheximide, ouabain and ethacrynic acid on various learning tasks in chickens. *Proceedings of the Royal Society of London A* 196, 171–195.

Rogers, L.J., Zappia, J.V. and Bullock, S.P. (1985) Testosterone and eye–brain asymmetry for copulation in chickens. *Experientia* 41, 1447 1449.

Rogers, L.J., Robinson, T. and Ehrlich, D. (1986) Role of the supraoptic decussation in the development of asymmetry of brain function in the chicken. *Developmental Brain Research* 28, 33–39.

Rogers, L.J., Adret, P. and Bolden, S.W. (1994a) Organisation of the thalamofugal visual projections in chick embryos, and a sex difference in light-stimulated development. *Experimental Brain Research* 97, 110–114.

Rogers, L.J., Ward, J.P. and Stafford, D. (1994b) Eye dominance in the small-eared bushbaby, *Otolemur garnettii*. *Neuropsychologica* 32, 257–264.

Romanoff, A.L. (1960) *The Avian Embryo*. Macmillan, New York.

Rose, S.P.R. (1989), Can memory be the brain's rosetta stone?. In: Cotterill, R.M.J. (ed.) *Models of Brain Function*. Cambridge University Press, Cambridge, pp. 1–13.

Rose, S.P.R. (1991a) Biochemical mechanisms involved in memory formation in the chick. In: Andrew R.J. (ed.) *Neural and Behavioural Plasticity: The Domestic Chick as a Model*. Oxford University Press, Oxford, pp. 277–304.

Rose, S.P.R. (1991b) How chicks make memories, the cellular cascade from c-fos to dendritic modelling. *Trends In Neurosciences* 14, 390–397.

Rose, S.P.R. (1992) *The Making of Memory: From Molecules to Mind*, Bantam Press, London.

Rose, S.P.R. and Csillag, A. (1985) Passive avoidance training results in lasting changes in deoxyglucose metabolism in left hemisphere regions of chick brain. *Behavioural and Neural Biology* 44, 315–324.

Rose, S.P.R., Gibbs, M.E. and Hambley, J.W. (1980) Transient increase in forebrain muscarinic cholinergic receptors following passive avoidance training. *Neuroscience* 5, 169–172.

Rostas, J. (1991) Molecular mechanisms of neuronal maturation: a model for synaptic plasticity. In: Andrew, R.J. (ed.) *Neural and Behavioural Plasticity: The Use of the Domestic Chick as a Model*. Oxford University Press, Oxford, pp. 177–211.

Rostas, J.A.P. and Jeffrey, P.L. (1981) Maturation of synapses in chicken forebrain. *Neuroscience Letters* 25, 299–304.

Rostas, J.A.P., Brent, V.A. and Guldner, F.H. (1984) The maturation of postsynaptic densities in chicken forebrain. *Neuroscience Letters* 45, 297–304.

Rostas, J.A.P., Kavanagh, J.M., Dodd, P.R., Heath, J.W. and Powis, D.A. (1991) Mechanisms of synaptic plasticity. *Molecular Neurobiology* 5, 203–216.

Rotenberg, V.S. (1992) Sleep and memory I: The influence of different sleep stages on memory. *Neuroscience and Biobehavioral Reviews* 16, 497–502.

Rubel, E.W. (1978) Ontogeny of structure and function in the vertebrate auditory system. In: Jacobson, M. (ed.) *Handbook of Sensory Physiology*, Vol. 9. Springer-Verlag, New York, pp. 135–237.

Rubel, E.W. and Parks, T.N. (1975) Organization and development of brain stem auditory nuclei of the chicken: tonotopic organization of N. magnocellularis and N. laminaris. *Journal of Comparative Neurology* 164, 411–434.

Rubel, E.W., Smith, D.J. and Miller, L.C. (1976) Organisation and development of the brain stem auditory nuclei of the chicken: ontogeny of n. magnocellularis and n. laminaris. *Journal of Comparative Neurology* 166, 469–490.

Rushen, J. (1984) Frequencies of agonistic behaviours as measures of aggression in chickens: a factor analysis. *Applied Animal Behaviour Science* 12, 167–176.

Russock, H.I. and Hale, E.B. (1979) Functional validation of the *Gallus* chicks' response to the maternal food call. *Zeitschrift für Tierpsychologie* 49, 250–259.

Sackeim, H.A., Gur, R.C. and Saucy, M.C. (1978) Emotions are expressed more intensely on the left side of the face. *Science* 202, 434–436.

Sagar, S.M., Sharp, F.R. and Curran, T. (1988) Expression of *c-fos* protein in brain: metabolic mapping at the cellular level. *Science* 240, 1328–1331.

Saleh, C.N. and Ehrlich, D. (1984) Composition of the supraoptic decussation in the chick (*Gallus gallus*). A possible fact limiting interhemispheric transfer of visual information. *Cell and Tissue Research* 236, 601–609.

Salk, L. (1962) Mothers' heartbeat as an imprinting stimulus. *Transcript of the New York Academy of Science* 24, 753–763.

Salzen, E.A. (1962) Imprinting and fear. *Zoological Society of London Symposium* 8, 199–217.

Salzen, E.A. (1966) Imprinting in birds and primates. *Behaviour* 28, 232–254.

Salzen, E.A. (1991) Learning paradigms and critical chick brain regions. *Abstracts of Avian Learning and Plasticity Conference*, Milton Keynes, UK, p. 17.

Salzen, E.A. and Cornell, J.M. (1969) Self perception and species recognition in birds. *Behaviour* 30, 44–65.

Salzen, E.A. and Meyer, C.C. (1968) Reversibility of imprinting. *Journal of Comparative and Physiological Psychology* 66, 269–275.

Salzen, E.A. and Sluckin, W. (1959) The incidence of the following response and the duration of responsiveness in domestic fowl. *Animal Behaviour* 11, 62–65.

Salzen, E.A., Parker, D.M. and Williamson, A.J. (1975) A forebrain lesion preventing imprinting in domestic chicks. *Experimental Brain Research* 24, 145–157.

Sanderson, C.A. and Rogers, L.J. (1981) 2,4,5-Trichlorophenoxyacetic acid causes behavioral effects in chickens at environmentally relevant doses. *Science* 211, 593–595.

Sandi, C., Rose, S.P.R. and Patterson, T.A. (1992) Unilateral hippocampal lesions prevent recall of a passive avoidance task in day-old chicks. *Neuroscience Letters* 141, 255–258.

Sandi, C., Patterson, T.A. and Rose, S.P.R. (1993) Visual input and lateralization of brain function in learning in the chick. *Neuroscience* 52, 393–401.

Saunders, J.C., Coles, R.B. and Gates, B.R. (1973) The development of auditory evoked responses in the cochlea and cochlear nuclei of the chick. *Brain Research* 63, 59–74.

Schaefer, H.H. and Hess, E.H. (1959) Color preferences in imprinting objects. *Zeitschrift für Tierpsychologie* 16, 161–172.

Schaeffel, F., Howland, H.C. and Farkas, L. (1986) Natural accommodation in the growing chicken. *Vision Research* 26, 1977–1993.

Schaeffel, F., Rohrer, B., Lemmer, T. and Zrenner, E. (1991) Diurnal control of rod function in the chicken. *Visual Neuroscience* 6, 641–653.

Scharff, C. and Nottebohm, F. (1991) A comparative study of the behavioral deficits following lesions of various parts of the zebra finch song system: implications for vocal learning. *Journal of Neuroscience* 11, 2896–2913.

Schifferli, A. (1948) Über Markscheidenbildung im Gehirn von Huhn und Star. *Revue Suisse* 455, 117–212.

Schneirla, T.C. (1965) Aspects of stimulation and organisation in approach/withdrawal processes underlying vertebrate behavioral development. In: Lehrman, D.S., Hinde, R.A. and Shaw, E. (eds) *Advances in the Study of Behavior*, vol. I, Academic Press, New York, pp. 1–74.

Schneirla, T.C. (1966) Behavioral development and comparative psychology. *Quarterly Review of Biology* 41, 283–302.

Schwabl, H. (1993) Yolk is a source of maternal testosterone for developing birds. *Proceedings of the National Academy of Sciences USA* 90, 11446–11450.

Schwarz, I.M. and Rogers, L.J. (1992) Testosterone: a role in the development of brain asymmetry in the chick. *Neuroscience Letters* 146, 167–170.

Sdraulig, R., Rogers, L.J. and Boura, A.L.A. (1980) Glutamate and specific perceptual input interact to cause retarded learning in chicks. *Physiology and Behavior* 24, 493–500.

Sedláček, J. (1964) Further findings on the conditions of formation of the temporary connection in chick embryos. *Physiologia Bohemoslovenica* 13, 411–420.

Sedláček, J. (1967) Development of optic evoked potentials in chick embryos. *Physiologia Bohemoslovenica* 13, 268–273.

Sedláček, J. (1972) Development of the optic afferent system in chick embryos. In: Newton, G. and Riesen, A. (eds) *Advances in Psychobiology*, vol. 1. Wiley, New York, pp. 129–170.

Seiler, N. and Sarhan, S. (1983) Metabolic routes of GABA formation in chick embryo brain. *Neurochemistry International* 5, 625–633.

Seller, T.J. (1979) Unilateral nervous control of the syrinx in Java sparrows. *Journal of Comparative Physiology A* 129, 281–288.

Seyfarth, R.M., Cheney, D.L. and Marler, P. (1980) Monkey responses to three different alarm calls: evidence for predator classification and semantic communication. *Science* 210, 801–803.

Shapiro, J.L. (1981) Pre-hatching influences that can potentially mediate posthatching attachments in birds. *Bird Behaviour* 3, 1–18.

Sherman, G.F. and Galaburda, A.M. (1985) Asymmetries in anatomy and pathology in the rodent brain. In: Glick S.D. (ed.) *Cerebral Lateralization in Nonhuman Species.* Academic Press, New York, pp. 185–231.

Sherry, D.F. (1977) Parental food calling and the role of the young in the Burmese red jungle fowl (*Gallus gallus spadiceus*). *Animal Behaviour* 25, 594–601.

Sherry, D.F. (1981) Parental care and the development of thermoregulation in red jungle fowl. *Behaviour* 76, 250–279.

Sherry, D.F. (1989) Food storing in the Paridae. *Wilson Bulletin* 10, 289–304.

Sherry, D.F., Vaccarino, A.L., Buckenham, K and Herz, R.S. (1989) The hippocampal complex of food-storing birds. *Brain, Behaviour and Evolution* 34, 308–317.

Sherry, D.F., Jacobs, L.F. and Gaulin, S.J.C. (1992) Spatial memory and adaptive specialization of the hippocampus. *Trends in Neurosciences* 15, 298–303.

Sherry, D.F., Forbes, M.R., Khurgel, M. and Ivy, G.O. (1993) Females have a larger hippocampus than males in the brood-parasitic brown-headed cowbird. *Proceedings of the National Academy of Sciences USA* 90, 7839–7843.

Shettleworth, S.J. (1990) Spatial memory in food-storing birds. *Philosophical Transactions of the Royal Society of London B* 329, 143–151.

Shimizu, T. and Hodos, W. (1989) Reversal learning in pigeon: effects of selective lesions of the Wulst. *Behavioral Neuroscience* 103, 262–272.

Shutze, J.V., Lauber, J.K., Kato, M. and Wilson, W.O. (1962) Influence of incandescent and coloured light on chicken embryos during incubation. *Nature* 10, 594–595.

Siegel, P.B., Isakson, S.T., Coleman, F.N. and Huffman, B.J. (1969) Photoacceleration of development in chick embryos. *Comparative Biochemistry and Physiology* 28, 753–758.

Siekevitz, P. (1985) The postsynaptic density: a possible role in long lasting effects in central nervous system. *Proceedings of the National Academy of Sciences USA* 82, 3494–3498.

Simner, M.L. (1966) Relationship between cardiac rate and vocal activity in newly hatched chicks. *Journal of Comparative and Physiological Psychology* 61, 496–498.

Simner, M.L. and Kaplan, W. (1977) The cardiac self-stimulation hypothesis and the chick's differential attraction toward intermittent sound. *Developmental Psychobiology* 10, 177–186.

Sisson, R.F. (1974) The heron that fishes with bait. *National Geographic* 45, 142–147.

Sluckin, W. (1966) *Imprinting and Early Learning.* Methuen, London. (First published in 1964.)

Sluckin, W. and Salzen, E.A. (1961) Imprinting and perceptual learning. *Quarterly Journal of Experimental Psychology* 13, 65–77.

Smith, F.V. and Bird, M.W. (1963) The relative attraction for the domestic chick of combinations of stimuli in different sensory modalities. *Animal Behaviour* 11, 300–305.

Smith, F.V. and Bird, M.W. (1964) The correlation of responsiveness to visual and auditory stimuli in the domestic chick. *Animal Behaviour* 12, 259–263.

Snapir, N., Robinzon, B. and Perek, M. (1973) Development of brain damage in the domestic fowl injected with monosodium glutamate at 5 days of age. *Pathologia Europoea* 8, 265–277.

Snow, D.W. (1961) The natural history of the oilbird, *Steatornis caripensis*, in Trinidad. Part I. General behavior and breeding habits. *Zoologica* 46, 27–48.

Sobue, K. and Nakajima, T. (1978). Changes in concentrations of polyamines and γ-aminobutyric acid and their formation in chick embryo brain during development. *Journal of Neurochemistry* 30, 277–279.

Soliman, K.F.A. and Huston, T.M. (1972) The photoelectric plethysmography technique for recording heart rate in chick embryos. *Poultry Science* 51, 651–654.

Solodkin, M., Cardona, A. and Corsi-Cabrera, M. (1985) Paradoxical sleep augmentation after imprinting in the domestic chick. *Physiology and Behaviour* 35, 343–348.

Somohano, F., Roberts, P.J. and López–Colomé, A.M. (1988) Maturational changes in retinal excitatory amino acid receptors. *Developmental Brain Research* 42, 59–67.

Spalding, D. (1873) Instinct. *MacMillan's Magazine* 27, 282–293. Reprinted 1954 *British Journal of Animal Behaviour* 2, 2–11.

Spencer, S.G. and Robinson, J.R. (1989) An autoradiographic analysis of neurogenesis in the chick retina in vitro and in vivo. *Neuroscience* 32, 801–812.

Spira, A.W., Millar, T.J., Ishimoto, I., Epstein, M.L., Johnson, C.D., Dahl, J.L. and Morgan, I. G. (1987) Localization of choline acetyltransferase-like immunoreactivity in the embryonic chick retina. *Journal of Comparative Neurology* 260, 526–538.

Squire, L.R., Weinberger, N.M., Lynch, G. and McGaugh, J.L. (1991) *Memory: Organization and Locus of Change*. Oxford University Press, Oxford.

Starke, K., Göthert, M. and Kilbinger, H. (1989) Modulation of neurotransmitter release by presynaptic autoreceptors. *Physiological Reviews* 69, 864–989.

Stattelman, A.J., Talbot, R.B. and Coulter, D.B. (1975) Olfactory thresholds in pigeons (*Columba livia*), quail (*Colinus virginianus*) and chickens (*Gallus gallus*). *Comparative Biochemistry and Physiology* 50A, 807–809.

Staubli, U. and Lynch, G. (1991) NMDA receptors and memory: evidence from pharmacological and correlational studies. In: Kozikowski, A.P. (ed.) *Neurobiology of the NMDA Receptor: From Chemistry to the Clinic*. VCH Publishers, New York, pp. 129–148.

Steele, R.J. and Stewart, M.G. (1993) 7-Chlorokynurenate, an antagonist of the glycine binding site on the NMDA receptor, inhibits memory formation in day-old chicks (*Gallus domesticus*). *Behavioral and Neural Biology* 60, 89–92.

Stevens, T.A. and Krebs, J.R. (1986) Retrieval of stored seeds by marsh tits *Parus palustris* in the field. *Ibis* 128, 513–525.

Stewart, M.G. and Bourne, R.C. (1986) Ontogeny of ^{3}H-muscimol binding to membranes of chick forebrain. *Experimental Brain Research* 243,1–4.

Stewart, M.G., Rose, S.P.R., King, T.S., Gabbot, P.L.A. and Bourne, R. (1984) Hemispheric asymmetry of synapses in the chick medial hyperstriatum ventrale following passive avoidance training: a stereological investigation. *Developmental Brain Research* 12, 261–269.

Stewart, M.G., Csillag, A. and Rose, S.P.R. (1987) Alterations in synaptic structure in the paleostriatal complex of the domestic chick, *Gallus domesticus*, following passive avoidance training. *Brain Research* 426, 69–81.

Stewart, M.G., Bourne, R.C., Chmielowska, J., Kalman, M., Csillag, A. and Stanford, D. (1988) Quantitative autoradiographic analysis of the distribution of [^{3}H]muscimol binding to GABA receptors in chick brain. *Brain Research* 456, 387–391.

Stewart, M.G., Bourne, R.C. and Steele, R.J. (1992a) Quantitative autoradiographic demonstration of changes in binding to NMDA sensitive ^{3}H-glutamate and ^{3}H-MK801, but not ^{3}H-AMPA receptors in chick forebrain 30 minutes after passive avoidance training. *European Journal of Neuroscience* 4, 936–943.

Stewart, M.G., Rogers, L.J., Davies, H.A. and Bolden, S.W. (1992b) Structural asymmetry in the thalamofugal visual projections in 2-day-old chick is correlated with a hemisphere difference in synaptic density in the hyperstriatum accessorium. *Brain Research* 585, 381–385.

Stone, N.D., Siegel, P.B., Adkisson, C.S. and Gross, W.B. (1984) Vocalizations and behavior of two commercial stocks of chickens. *Poultry Science* 63, 616–619.

Stone, W.S. and Gold, P.E. (1988) Sleep and memory relationships in intact old and amnestic young rats. *Neurobiology of Aging* 9, 719–727.

Stone, W.S., Walker, D.L. and Gold, P.E. (1992) Sleep deficits in rats after NMDA receptor blockade. *Physiology and Behavior* 52, 609–612.

Suboski, M.D. (1987) Environmental variables and releasing-valence transfer in stimulus-directed pecking of chicks. *Behavioural and Neural Biology* 47, 262–274.

Suboski, M.D. and Bartashumas, C. (1984) Mechanisms for social transmission of pecking preferences to neonatal chicks. *Journal of Experimental Psychology* 10, 189–192.

Takamatsu, K. and Tsukada, Y. (1985) Neurochemical studies on imprinting behavior in chick and duckling. *Neurochemical Research* 10, 1371–1391.

Tamarin, A., Crawley, A., Lee, J. and Tickle, C. (1984) Analysis of upper beak defects in chicken embryos following treatment with retinoic acid. *Journal of Embryological and Experimental Morphology* 84, 105–123.

Tanabe, Y., Nakamura, T., Fujioka, K. and Doi, O. (1979) Production and secretion of sex steroid hormones by the testes, the ovary and the adrenal glands of embryonic and young chickens (*Gallus domesticus*). *General and Comparative Endocrinology* 39, 26–33.

Tauber, H., Waehneldt, T.V. and Neuhoff, V. (1980) Myelination in rabbit optic nerves is accelerated by artificial eye opening. *Neuroscience Letters* 16, 239–244.

Tauson, R. and Svensson, S.A. (1980) Influence of plumage condition on the hen's feed requirement. *Swedish Journal of Agricultural Research* 10, 35–39.

Ten Cate, C. (1989a) Stimulus movement, hen behaviour and filial imprinting in Japanese quail (*Coturnix coturnix japonica*). *Ethology* 82, 287–306.

Ten Cate, C. (1989b) Behavioral devleopment: toward understanding processes. In: Bateson, P.P.G. and Klopfer, P. (ed.) *Perspectives in Ethology*, Plenum Press, New York, pp. 243–269.

Terrace, H.S. (1985) Animal cognition: thinking without language. *Philosophical Transactions of the Royal Society of London*, B. 308, 113–128.

Thanos, S. and Bonhoeffer, F. (1983) Investigations on development and topographic order of retinotectal axons: anterograde and retrograde staining of axons and their perikarya with rhodamine in vivo. *Journal of Comparative Neurology* 219, 420–430.

Thanos, S. and Bonhoeffer, F. (1987) Axonal arborisation in the developing chick retinotectal system. *Journal of Comparative Neurology* 261, 155–164.

Thiriet, G., Kempf, J. and Ebel, A. (1992) Distribution of cholinergic neurones in the chick spinal cord during embryonic development. Comparison of ChAT immunocytochemistry with AChE histochemistry. *International Journal of Neuroscience* 10, 459–466.

Tolhurst, B.E.and Vince, M.A. (1976) Sensitivity to odours in the embryo of the domestic fowl. *Animal Behaviour* 24, 772–779.

Tolman, C.W. (1964) Social facilitation of feeding behaviour in the domestic chick. *Animal Behaviour* 12, 245–251.

Tolman, C.W. (1967a) The feeding behaviour of domestic chicks as a function of rate of pecking by a surrogate companion. *Behaviour* 29, 57–62.

Tolman, C.W. (1967b) The effect of tapping sounds on feeding behaviour of domestic chicks. *Animal Behaviour* 15, 145–148.

Tolman, C.W. (1968) The varieties of social stimulation in the feeding behaviour of domestic chicks. *Behaviour* 30, 275–286.

Tolman, C.W. and Wilson, G.F. (1965) Social feeding in domestic chicks. *Animal Behaviour* 13, 134–142.

Tsai, H.M., Garber, B.B. and Larramendi, L.M.H. (1981) [^{3}H]Thymidine autoradiographic analysis of telencephalic histogenesis in the chick embryo: I. Neuronal birthdates of telencephalic compartments *in situ*. *Journal of Comparative Neurology* 198, 275–292.

Tschanz, B. (1968) Trottellumen (*Uria aalge*). Die Entstehung der persönlichen Beziehung zwischen Jungvogel und Eltern. *Zeitschrift für Tierpsychologie* 4, 1–103.

Tuculescu, R.A. and Griswold, J.G. (1983) Prehatching interactions in domestic chicks. *Animal Behaviour* 31, 1–10.

Turner, E.R.A. (1964) Social feeding in birds. *Behaviour* 24, 1–45.

Uchiyama, H. (1989) Centrifugal pathways to the retina: influence of the optic tectum. *Visual Neuroscience* 3, 183–206.

Vallortigara, G. (1989) Behavioral asymmetries in visual learning of young chickens. *Physiology and Behavior* 44, 797–800.

Vallortigara, G. (1992a) Affiliation and aggression as related to gender in domestic chicks (*Gallus gallus*). *Journal of Comparative Psychology* 106, 53–57.

Vallortigara, G. (1992b) Right hemisphere advantage for social recognition in the chick. *Neuropsychologia* 30, 761–768.

Vallortigara, G. and Andrew, R.J. (1991) Lateralization of response by chicks to change in a model partner. *Animal Behaviour* 41, 187–194.

Vallortigara, G. and Andrew, R.J. (1994) Olfactory lateralization in the chick. *Neuropsychologia* 32, 417–423.

Vallortigara, G. and Zanforlin, M. (1988) Open-field behavior of young chicks (*Gallus gallus*): antipredatory responses, social reinstatement motivation, and gender effects. *Animal Learning and Behavior* 16, 359–362.

Vallortigara, G., Zanforlin, M. and Cailotto, M. (1988) Right–left asymmetry in position learning of male chicks. *Behavioural Brain Research* 27, 189–191.

Vallortigara, G., Cailotto, M. and Zanforlin, M. (1990) Sex differences in social reinstatement motivation of the domestic chick (*Gallus gallus*) revealed by runway tests with social and nonsocial reinforcement. *Journal of Comparative Psychology* 104, 361–367.

Vallortigara, G., Regolin, L. and Zanforlin, M. (1994) The development of responses to novel-coloured objects in male and female domestic chicks. *Behavioural Processes* 31, 219–230.

van Kampen, H.S. (1993a) An analysis of the learning process underlying filial imprinting. PhD thesis, University of Groningen, The Netherlands.

van Kampen, H.S. (1993b) Filial imprinting and associative learning: similar mechanisms? *Netherlands Journal of Zoology* 43, 143–154.

van Kampen, H.S. and Bolhuis, J.J. (1991) Auditory learning and filial imprinting in the chick. *Behaviour* 117, 303–319.

van Kampen, H.S. and Bolhuis, J.J. (1993) Interaction between auditory and visual learning during imprinting. *Animal Behaviour* 45, 623–625.

van Kampen, H.S. and de Vos, G.J. (1991) Learning about the shape of an imprinting object varies with its colour. *Animal Behaviour* 42, 328–329.

van Kampen, H.S. and de Vos, G.J. (1992) Memory for the spatial position of an imprinting object in junglefowl chicks. *Behaviour* 122, 26–40.

van Kampen, H.S., de Haan, J. and de Vos, G.J. (1994) Potentiation in learning about the visual features of an imprinting stimulus. *Animal Behaviour* 47, 1468–1470.

Vanzulli, A. and Garcia-Austt, E. (1963) Development of cochlear microphonic potentials in the chick embryo. *Acta Neurologica Latinoamericana* 9, 19–23.

Vaughan, W., and Greene, S.L. (1984) Pigeon visual memory capacity. *Journal of Experimental Psychology: Animal Behaviour Processes* 10, 256–271.

Vazquez-Nin, G.H. and Sotelo, J.R. (1968) Electron microscope study of the developing nerve terminals in the acoustic organs of the chick embryo. *Zeitschrift für Zellforschung und mikroskopische Anatomie* 92, 325–338.

Vestergaard, K.S., Kruijt, J.P. and Hogan, J.A. (1993) Feather pecking and chronic fear in groups of red junglefowl: their relations to dustbathing, rearing environment and social status. *Animal Behaviour* 45, 1127–1140.

Vidal, J.M. (1980) The relations between filial and sexual imprinting in the domestic fowl: effects of age and social experience. *Animal Behaviour* 28, 880–891.

Vince, M.A. (1964) Synchronisation of hatching in the bobwhite quail (*Colinus virginianus*). *Nature* 203, 1192–1193.

Vince, M.A. (1966) Potential stimulation produced by avian embryos. *Animal Behaviour* 14, 34–40.

Vince, M.A. (1970) Some aspects of hatching behaviour. In: Freeman, B.M. and Gordon, R.F. (eds) *Aspects of Poultry Behaviour*. British Poultry Science Ltd, Edinburgh, pp. 33–62.

Vince, M.A. (1973) Some environmental effects on the activity and development of the avian embryo. In: Gottlieb, G. (ed.) *Behavioural Embryology*. Academic Press, New York, pp. 285–323.

Vince, M.A. (1977) Taste sensitivity in the embryo of the domestic fowl. *Animal Behaviour* 25, 797–805.

Vince, M.A. (1980) The posthatching consequences of prehatching stimulation: changes with amount of prehatching and posthatching exposure. *Behaviour* 75, 36–53.

Vince, M.A. and Cheng, R. (1970) Effects of stimulation on the duration of lung ventilation in quail fetuses. *Journal of Experimental Zoology* 175, 477–486.

Vince, M.A. and Chinn, S. (1972) Effects of external stimulation on the domestic chick's capacity to stand and walk. *British Journal of Psychology* 63, 89–99.

Vince, M.A., Green, J. and Chinn, S. (1970) Acceleration of hatching in the domestic fowl. *British Poultry Science* 11, 483–488.

Vince, M.A., Reader, M. and Tolhurst, B. (1976) Effects of stimulation on embryonic activity in the chick. *Journal of Comparative and Physiological Psychology* 90, 221–230.

Vockel, A., Pröve, E. and Balthazart, J. (1990) Sex- and age-related differences in the activity of testosterone-metabolizing enzymes in microdissected nuclei of the zebra finch brain. *Brain Research* 511, 291–302.

von Bartheld, C.S., Young, W. and Rubel, E.W. (1992) Normal and abnormal pathfinding of facial nerve fibres in the chick embryo. *Journal of Neurobiology* 23, 1021–1036.

von Fersen, L. and Güntürkün, O. (1990) Visual memory lateralization in pigeons. *Neuropsychologia* 28, 1–7.

Voukelatou, G., Aletras, A.J., Tsourinakis, T. and Kouvelas, E.D. (1992) Glutamate-like immunoreactivity in chick cerebellum and optic tectum. *Neurochemical Research* 17, 1267–1273.

Wallhäusser, E. and Scheich, H. (1987) Auditory imprinting leads to differential 2-deoxyglucose uptake and dendritic spine loss in the chick rostral forebrain. *Developmental Brain Research* 31, 29–44.

Wallman, J. and Pettigrew, J.D. (1985) Conjugate and disjunctive saccades in two avian species with contrasting oculomotor strategies. *Journal of Neuroscience* 5, 1418–1428.

Walls, G.L. (1942) *The Vertebrate Eye and its Adaptive Radiation*. Hafner, New York.

Ward, J.P. and Hopkins, W.D. (1993) *Primate Laterality: Current Behavioral Evidence of*

Primate Asymmetries, Springer-Verlag, New York.

Watanabe, S., Lea, S.E.G. and Dittrich, W. (1993) What can we learn from experiments on pigeon concept discrimination? In: Zeigler, H.P. and Bischof, H-J. (eds) *Vision, Brain, and Behavior in Birds*. MIT Press, Cambridge, MA, pp. 351–376.

Weidmann, U. (1958) Verhaltensstudien an der Stockente. II. Versuche zur Auslösung und Prägung der Nachfolge- und Anschlussreaktion. *Zeitschrift für Tierpsychologie* 15, 277–300.

Weidner, C., Repérant, J., Desroches, A-M., Micelli, D. and Vesselkin, N.P. (1987) Nuclear origin of the centrifugal visual pathway in birds of prey. *Brain Research* 436, 153–160.

Weiskrantz, L. (1990) Problems of learning and memory: one or multiple memory systems? *Philosophical Transactions of the Royal Society of London B* 329, 99–108.

Wendel Yee, G. and Abbott, U.K. (1978) Facial development in normal and mutant chick embryos: scanning electron microscopy of primary palate formation. *Journal of Experimental Zoology* 206, 307–322.

Wenzel, B.M. (1968) Olfactory prowess of the kiwi. *Nature* 220, 1133–1134.

Wiesel, T.N. and Hubel, D.H. (1963) Single-cell responses in striate cortex of kittens deprived of vision in one eye. *Journal of Neurophysiology* 26, 1003–1017.

Will, L.A. and Meller, S.M. (1981) Primary palatal development in the chick. *Journal of Morphology* 169, 185–190.

Williams, H., Crane, L.A., Hale, T.K., Espositeo, M.A. and Nottebohm, F. (1992) Right-side dominance for song control in the zebra finch. *Journal of Neurobiology* 23, 1006–1020.

Wilson, J.X., Lui, E.M.K. and del Maestro, R.F. (1992) Development profiles of antioxidant enzymes and trace metals in chick embryo. *Mechanisms of Ageing and Development* 65, 51–64.

Wilson, P. (1980) The organisation of the visual hyperstriatum in the domestic chick. II. Receptive field properties of single units. *Brain Research* 188, 333–345.

Winkler, D.W. (1993) Testosterone in egg yolks: an ornithologist's perspective. *Proceedings of the National Academy of the Sciences USA* 90, 11439–11441.

Winson, J. (1993) The biology and function of rapid eye movement sleep. *Current Opinion in Neurobiology* 3, 243–248.

Wolff, J.R. (1981) Evidence for a dual role of GABA as a synaptic transmitter and a promotor of synaptogenesis. In: De Feudis, F.V. and Mandel, P. (eds) *Amino Acid Neurotransmitters*. Raven Press, New York, pp. 459–465.

Wood-Gush, D.G.M. (1956) The agonistic and courtship behaviour of the Brown Leghorn cock. *British Journal of Animal Behaviour* 4, 133–142.

Wood-Gush, D.G.M. (1971) *The Behaviour of the Domestic Fowl*, Heinemann, London.

Wood-Gush, D.G.M. (1983) *Elements of Ethology*, Chapman and Hall, London.

Woods, J.E. and Brazzill, D.M. (1981) Plasma 17β-estradiol levels in the chick embryo. *General and Comparative Endocrinology* 44, 37–43.

Woods, J.E., Simpson, R.M. and Moore, P.L. (1975) Plasma testosterone levels in the chick embryo. *General and Comparative Endocrinology* 27, 543–547.

Workman, L. and Andrew, R.J. (1986) Asymmetries of eye use in birds. *Animal Behaviour* 34, 1582–1584.

Workman, L. and Andrew, R.J. (1989) Simultaneous changes in behaviour and lateralization during the development of male and female domestic chicks. *Animal*

Behaviour 38, 596–605.

Workman, L. and Rogers, L.J. (1990) Pecking preferences in young chicks: effects of nutritive reward and beak-trimming. *Applied Animal Behaviour Science* 26, 115–126.

Workman, L., Kent, J.P. and Andrew, R.J. (1991) Development of behaviour in the chick. In: Andrew, R.J. (ed.) *Neural and Behavioural Plasticity: the Use of the Domestic Chick as a Model.* Oxford University Press, Oxford, pp. 166–173.

Young, C.E. and Rogers, L.J. (1978) Effects of steroidal hormones on sexual, attack, and search behavior in the isolated male chick. *Hormones and Behavior* 10, 107–117.

Youngren, O.M., Peek, F.W. and Phillips, R.E. (1974) Repetitive vocalisations evoked by local electrical stimulation of avian brains. III. Evoked activity in the tracheal muscles of the chicken (*Gallus gallus*). *Brain Behavior and Evolution* 9, 393–421.

Zajonc, R.B., Wilson, W.R. and Rajecki, D.W. (1975) Affiliation and social discrimination produced by brief exposure in day-old domestic chicks. *Animal Behaviour* 23, 131–138.

Zappia, J.V. and Rogers, L.J. (1983) Light experience during development affects asymmetry of forebrain function in chickens. *Developmental Brain Research* 11, 93–106.

Zappia, J.V. and Rogers, L.J. (1987) Sex differences and reversal of brain asymmetry by testosterone in chickens. *Behavioural Brain Research* 23, 261–267.

Zayan, R. (1987) An analysis of dominance and subordination experiences in sequences of paired encounters between hens. In: Zayan, R. and Duncan, I.J.H. (eds) *Cognitive Aspects of Social Behaviour in the Domestic Fowl.* Elsevier, Amsterdam, pp. 182–320.

Zayan, R. and Duncan, I.J.H. (1987) *Cognitive Aspects of Social Behaviour in the Domestic Fowl.* Elsevier, Amsterdam.

INDEX